Created in cooperation
with Fluke Corporation

Motor and Drive Troubleshooting: Basic Testing to Advanced Diagnostics

AMERICAN TECHNICAL PUBLISHERS
ORLAND PARK, ILLINOIS 60467-5756

Glen A. Mazur

Motor and Drive Troubleshooting: Basic Testing to Advanced Diagnostics contains procedures commonly practiced in industry and the trade. Specific procedures vary with each task and must be performed by a qualified person. For maximum safety, always refer to specific manufacturer recommendations, insurance regulations, specific job site and plant procedures, applicable federal, state, and local regulations, and any authority having jurisdiction. The material contained is intended to be an educational resource for the user. Neither American Technical Publishers, Inc. nor the Fluke Corporation is liable for any claims, losses, or damages, including property damage or personal injury, incurred by reliance on this information.

American Technical Publishers, Inc., Editorial Staff

Editor in Chief:
    Jonathan F. Gosse
Vice President—Production:
    Peter A. Zurlis
Art Manager:
    James M. Clarke
Technical Editor:
    James T. Gresens
Copy Editor:
    Jeana M. Platz
Cover Design:
    Jennifer M. Hines
Illustration/Layout:
    Thomas E. Zabinski

1 2 3 4 5 6 7 8 9 – 11 – 9 8 7 6 5 4

Printed in the United States of America

ISBN 978-0-8269-1538-2

 This book is printed on recycled paper.

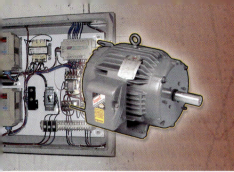

Contents

Motor and Drive Troubleshooting: Basic Testing to Advanced Diagnostics is designed to provide an introduction to portable oscilloscope operation principles and procedures. Portable oscilloscopes have become predictive and preventive maintenance diagnostic tools for technicians in industry. *Motor and Drive Troubleshooting: Basic Testing to Advanced Diagnostics* covers the fundamentals of using portable oscilloscopes to detect and analyze electric motor and drive problems that can cause system or process downtime. Safety, inspection procedures, analysis, reporting, and documentation are detailed. Tech-Tips included relate to portable oscilloscope safety practices, troubleshooting tips, applications, and selected chapter content.

Additional information related to testing, troubleshooting, and maintenance principles is available in other ATP learning materials. To obtain information on related training products, please visit the ATP web site at www.go2atp.com or call 1-800-323-3471.

Motor and Drive Principles

Electric motors are used to produce more work than any other electrical component. They consume more than 65% of all electricity produced. Electric motors range in output capacity from a few milliwatts to thousands of kilowatts. Because electric motors perform essential work and consume large amounts of power, they must be controlled in the most productive and efficient manner possible.

Understanding how motors operate, how to apply them in specific applications, and how best to control them assures that they can perform the required work. Understanding what test instruments to use when installing, maintaining, and troubleshooting them assures that they can safely operate for extended time periods with minimal downtime.

ELECTRIC MOTORS

Electric motors are available in different sizes and types to meet the different requirements of applications requiring a rotating force. The type of motor used depends upon the available power (DC or AC), type of load to be driven, environmental conditions, and efficiency. Electric motors are rated for power, voltage, current, frequency, and speed. Understanding changes in load voltage, current, frequency, and speed helps determine the type of motor to use for the application. Additionally, the proper motor insulation should be used for the application.

A *direct current (DC) motor* is a motor that uses direct current connected to the field windings and rotating armature to produce rotation. The four basic types of DC motors are series, shunt, compound, and permanent-magnet motors. The advantage of using DC motors is that they can be used in portable applications and motor speed is easily controlled by changing the applied voltage. DC motors are often used for portable applications, but AC motors are replacing DC motors in most other applications.

An *alternating current (AC) motor* is a motor that uses alternating current connected to a stator (stationary windings) to produce a force on a rotor through a magnetic field. The advantage of an AC motor is that there is less maintenance required because there are no brushes to maintain. The two main types of AC motors are single-phase (1ϕ) and three-phase (3ϕ). Single-phase motors include shaded-pole, split-phase, capacitor-start, capacitor-run, and capacitor start-and-run. Single-phase motors are primarily used when there is no three-phase power available.

However, because motor drives are available that convert single-phase into three-phase for driving three-phase motors, three-phase motors are replacing single-phase motors in HVAC and other applications that once used single-phase motors. Three-phase motors are more energy efficient and have no capacitors or centrifugal switches to maintain. Three-phase motors have commonly been the standard motor for most applications requiring a motor over 1 HP (746 W) and are available in fractional HP sizes to replace single-phase motors. Three-phase motors are the most common motor type controlled by a motor drive.

Motor Power Ratings

An *electric motor* is a machine that converts electrical power into rotating mechanical force (torque) on a shaft, which is used to produce work. The amount of power that a motor produces is expressed in watts (W), kilowatts (kW), or horsepower (HP). The motor power rating is located on the motor's nameplate, which is affixed to the motor. Motor nameplates also include information such as voltage and RPM ratings. **See Figure 1-1.** For conversion purposes, 746 W is equal to 1 HP. If a motor is rated in kW, a comparison to HP can be calculated by multiplying the kW rating of the motor by 1.34 to get the equivalent HP rating.

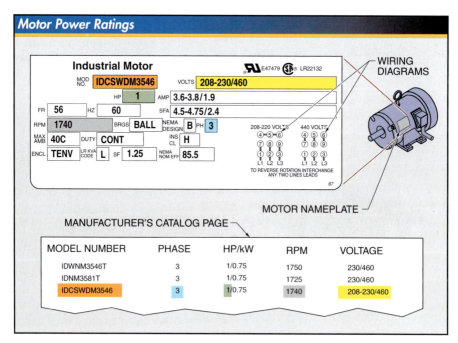

Figure 1-1. The motor power rating is located on the motor's nameplate, which is affixed to the motor, and also includes information such as voltage and RPM ratings.

A *portable oscilloscope* (Scope-Meter™) is a test instrument that measures and displays the waveforms of high-voltage power, low-voltage control, and digital signals. When a test instrument such as a portable oscilloscope is used for testing electric motors, it can be used to measure the different types of power that pertain to the motor. For example, a portable oscilloscope can be used to measure true power, apparent power, reactive power, or power factor. Multichannel portable oscilloscopes can measure multiple signals simultaneously (e.g., all three phases of a variable frequency drive) and different aspects of the same signal simultaneously (e.g., voltage and current). **See Figure 1-2.**

True power (P_T) is the actual power used in an electrical circuit and is measured in watts (W) or kilowatts (kW). Electric utility companies charge their customers based on the amount of true power used. *Apparent power (P_A)* is the product of voltage and current in a circuit calculated without considering the phase shift that may be present between the voltage and current in a circuit and is measured in volt amps (VA) or kilovolt amps (kVA). *Reactive power (VAR)* is power supplied to reactive loads, such as motor coil windings. Reactive power is measured in volt-amps reactive (VAR).

TECH TIP

Utility companies penalize customers that have low power factors. Facilities with many inductive loads, such as motor windings, solenoids, or transformers, have a poor power factor because voltage is leading the current.

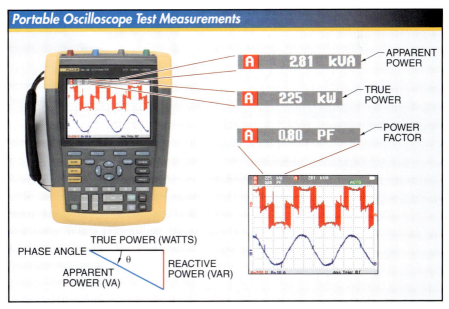

Portable Oscilloscope Test Measurements

Figure 1-2. A portable oscilloscope can be used to measure true power, apparent power, reactive power, or power factor.

Power factor (PF) is the ratio of true power used in an AC circuit to apparent power used by or delivered to the circuit. Power factor is expressed as a percentage. True power equals apparent power only when the power factor is 100% or 1. When the power factor is less than 100%, the circuit is less efficient and has a higher operating cost because not all the current is performing work. Power sources, such as transformers, are rated in apparent power (kVA) since they must deliver all the power regardless of the load's power factor.

Motor Voltage Ratings

All motors are designed for optimum performance at a specific voltage level. However, not all power supplies can deliver the specified voltage at all times. Any increase or decrease in the voltage applied to a motor affects motor performance.

As the voltage applied to a motor decreases, motor torque and efficiency decrease and motor current and power factor increase. As the voltage increases, motor efficiency and power factor decrease and motor current and torque increase. The applied voltage to a motor should be within −10% to +5% of its voltage rating for best performance. **See Figure 1-3.**

TECH TIP

The power factor (PF) of a motor is lowest when the motor is not loaded and highest when the motor is loaded 50% to full-load.

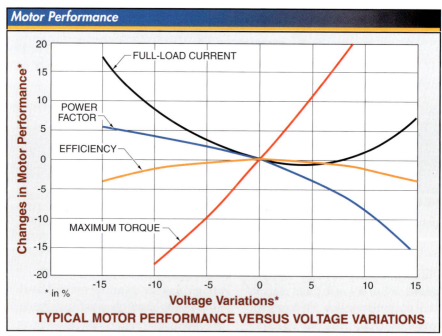

Figure 1-3. The applied voltage to a motor should be within −10% to +5% of its voltage rating for best performance.

Motor Current Ratings

All motors draw current in order to produce power. Current changes with a change in motor load. As the load on a motor increases, the current draw of the motor also increases. As the load on the motor decreases, the current draw of the motor also decreases. The nameplate rated current of a motor is the amount of current the motor draws when fully loaded. **See Figure 1-4.**

A motor attempts to drive a load, even if the load exceeds the motor's power rating. A motor's service factor rating indicates whether the motor can safely withstand an overload condition. A nameplate service rating of 1 (or no listing) indicates the motor is not designed to safely withstand an overloaded condition above the motor's rated power. A service factor of 1.15 indicates that the motor can safely withstand an overload 15% higher than the motors rated power. The International Electrotechnical Commission (IEC) uses eight designations (S1 through S8) to describe the duty cycles of electric motors. For example, S1 designates "Continuous Duty" and S8 designates "Continuous Operation with Periodic Changes in Load and Speed."

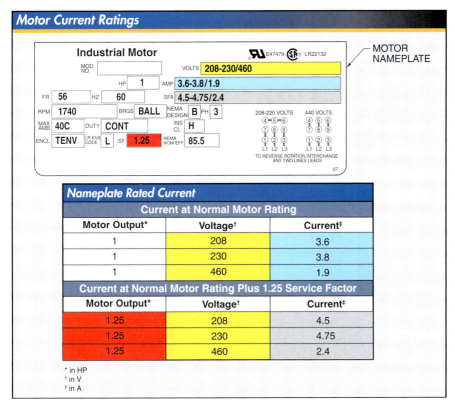

Figure 1-4. The nameplate rated current of a motor is the amount of current the motor will draw when fully loaded.

In order to deliver the higher power (*P*), the motor must draw more current (*I*) since the voltage (*E*) remains constant at the motor. The power formula states that $P = E \times I$. Therefore, for a motor to produce more power, the motor must draw more current. *Service factor amperage (SFA)* is the maximum current rating a motor can safely draw. The SFA is listed on an electric motor nameplate. A motor that draws more current than the SFA can become overloaded and be easily damaged, usually because of insulation breakdown.

Baldor Electric Co.

Motor ratings are listed on the motor nameplate, which is affixed directly to the motor.

Voltage and Current Changes

In most circuits, voltage remains within an acceptable range (−10% to +5%) of the nameplate rating, unless there is a problem. However, current constantly changes as loads are switched on and off and motor loads change. When there appears to be a problem in a circuit due to a low-voltage condition, such as computers resetting or lamps flickering, a portable oscilloscope can help determine the type of problem present and when it occurs. When a problem occurs, the two main conditions to observe are the relationship between when there is a change in the voltage and the direction the current changes (decreases or increases) at that time.

Voltage and Current Decreases. When voltage and current decrease together, the problem is in the power source and distribution system. The circuit is not overloaded. With a portable oscilloscope, data is recorded and can be viewed to help analyze the circuit. For example, with a 14.5% voltage decrease, current also decreases. The decrease in both voltage and current also causes a power decrease at electrical loads such as motors, lamps, and heating elements. **See Figure 1-5.**

A technician must realize when a problem occurs in order to better understand the problem and take corrective action. With a portable oscilloscope, the data can be analyzed by placing cursors at the start and end of the problem. With a multichannel oscilloscope, up to four signals can be monitored at once. This allows for simultaneous monitoring of changes in both voltage and current for more than one signal.

TECH TIP

With a handheld portable oscilloscope, each measurement can be enlarged with the "zoom" function, allowing technicians to closely view the waveform for slight differences in readings that can be an indication of a problem.

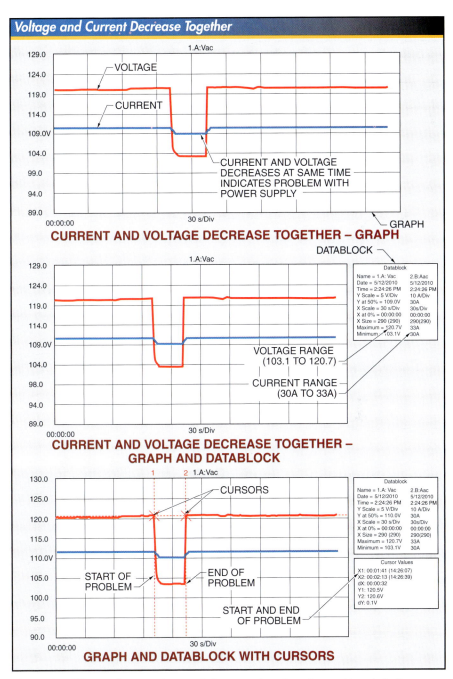

Figure 1-5. When voltage and current decrease together, the problem is in the power source and distribution system.

Voltage Decreases with Current Increases. When loads are switched on and motors require more power, current is increased. If the power source is not overloaded, the increase in current should not lower the voltage. For example, in a circuit where the current increases from 33 A to 152 A, the voltage decreases but remains within the acceptable operating range. **See Figure 1-6.**

Motor Frequency and Speed

Line frequency is the amount of complete electrical cycles per second of a power source. A cycle is one complete wave of alternating voltage or current. An alternation is one-half of a cycle. Period (*T*) is the time required to produce one complete cycle of a waveform. The line frequency rating of an electric motor is abbreviated on the nameplate as CY or CYC for cycle or Hz for hertz. *Hertz (Hz)* is the international unit

of frequency and is equal to cycles per second.

In the United States, 60 Hz is the standard power source line frequency. Canada, Mexico, and most of the Caribbean (Bahamas and Cayman Islands, for example) also use 60 Hz as the standard power source line frequency. All other countries use the 50 Hz frequency. Developing countries (Saudi Arabia, Colombia, Costa Rica, etc.) and countries using large numbers of U.S.-manufactured electrical products are changing their operating frequencies from 50 Hz to 60 Hz.

AC motors can have a 50 Hz, 60 Hz, or 50/60 Hz frequency rating. Changing the frequency changes the motor speed. Increasing the frequency increases motor speed, and decreasing the frequency decreases motor speed. For example, a motor with a 60 Hz nameplate rating operates at 75% speed at 45 Hz, 50% speed at 30 Hz, and 25% speed at 15 Hz.

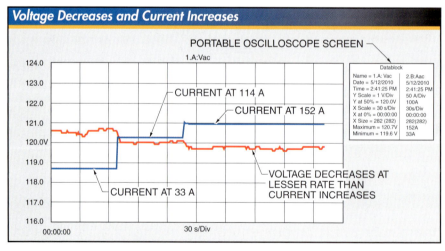

Figure 1-6. As current in the circuit increases, voltage decreases at a lesser rate.

A *motor drive* is an electronic unit designed to control the speed of a motor using solid-state devices. With modern motor drives, a motor that once could only be operated at the nameplate rated speed can now operate at speeds above and below the nameplate rated speed. Although a motor drive can operate a motor above its rated speed, it is usually unsafe to do so. Operating a motor above the speed at which the motor is designed to safely operate can cause the motor to break apart while in operation, creating a hazardous environment.

Most motor manufacturers balance their motors at speeds 25% over nameplate rated speed. Because of this, it is recommended to not operate a motor at 20% or more than its nameplate rated speed. To assure that the maximum frequency is within +20% or less of the motor's nameplate rating, the "maximum frequency" parameter should be programmed to no more than 60 Hz for a 50 Hz nameplate rated motor and 72 Hz for a 60 Hz nameplate rated motor. **See Figure 1-7.**

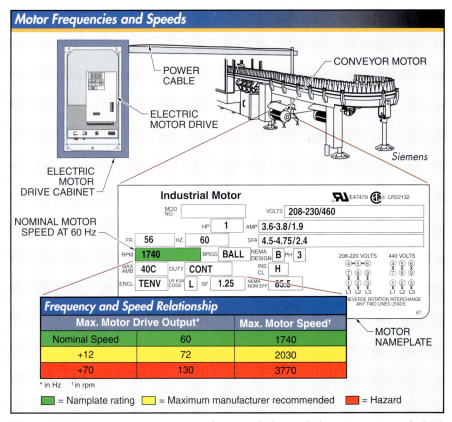

Figure 1-7. Because most motor manufacturers balance their motors at speeds 25% over nameplate rated speed, it is recommended to not operate a motor at 20% or more than its nameplate rated speed.

Motor drives can be set and programmed using either switches and potentiometers or a keypad. For example, a motor drive can use two-position switches and potentiometers to set the motor's operating conditions. The motor drive allows the maximum frequency output to be set, which also sets the motor's maximum operating speed. To set a motor drive's maximum frequency output, apply the following procedure:

1. Set the two-position switch to either the 50 Hz or 60 Hz position, depending upon the motor frequency nameplate listing.

2. Set the high-frequency potentiometer to the "nominal" position to limit the drive's frequency output to 50 Hz or 60 Hz based on the switch position.

3. Set the high-frequency potentiometer to a maximum of an additional 70 Hz from the nominal position.

Although motor drives can be set and programmed using switches and potentiometers, most motor drives use a keypad to program the drive's motor operating parameters. **See Figure 1-8.** To program a motor drive's maximum frequency output using a keypad, apply the following procedure:

1. Identify the parameter number used by the drive manufacturer to set the drive's maximum frequency output (located in the manufacturer's operating manual).

2. Enter the program and set the maximum drive output frequency. *Note:* When entering this parameter, the "Factory Default" setting (or last change entered) is displayed.

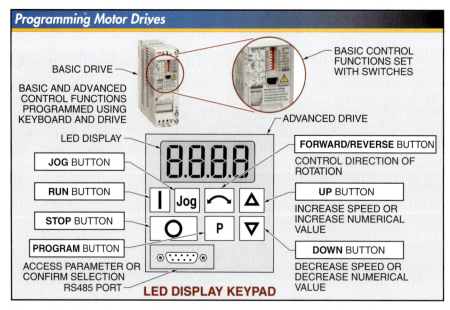

Figure 1-8. Although motor drives can be set and programmed using switches and potentiometers, most motor drives use a keypad to program motor operating parameters.

Caution: Although the drive can be set for a higher frequency output than it is rated for, limit the setting to no more than 20% above the motor's nameplate Hz rating. Operating a motor above the speed at which the motor is designed to operate safely can cause the motor to break apart while in operation, creating a hazardous situation.

Motor Frequency and Voltage

The voltage applied to the stator of an AC motor must be decreased by approximately the same amount as the frequency when controlling motor speed. Motor windings can quickly overheat, causing damage to the motor if voltage is not reduced as frequency is reduced. The *volts/hertz (V/Hz) ratio* is the ratio of the voltage and frequency applied to a motor.

The V/Hz ratio is calculated by dividing the rated nameplate voltage by the rated nameplate frequency. For example, the V/Hz ratio for a 230 V/50 Hz rated motor is 4.6 V/Hz (230 ÷ 50 = 4.6), a 230 V/60 Hz rated motor is 3.83 V/Hz, a 460 V/50 Hz rated motor is 9.2 V/Hz, and a 460 V/60 Hz rated motor is 7.67 V/Hz.

Boost covers two separate but closely related parameters: start boost and continuous boost. Certain electric motor drive applications require extra starting torque at low speeds, while other applications require extra torque to reach base speed. Start boost provides extra torque at startup by initially applying a higher voltage. Continuous boost provides extra torque by applying a higher voltage to reach base speed, with voltage not exceeding motor nameplate voltage. Both parameters alter the V/Hz curve to provide more torque and current. **See Figure 1-9.**

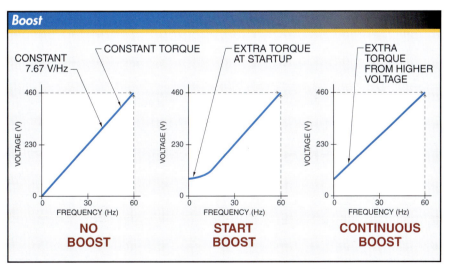

Figure 1-9. Start boost applies higher voltage initially and then follows the constant torque V/Hz pattern. Continuous boost applies higher voltage from zero to base speed.

Start boost and continuous boost are measured in volts or percent of rated motor current. There is some interaction between the two parameters. The default settings vary among electric motor drive manufacturers.

TECH TIP

In addition to portable oscilloscopes, thermal imagers can be used to perform motor circuit analysis (MCA).

Above approximately 15 Hz, the amount of voltage required to keep the V/Hz ratio linear is a constant value. Below 15 Hz, the voltage applied to a motor stator must be boosted to compensate for the large power loss AC motors experience at low speeds. The amount of voltage boost depends on the specific motor and the type of load for which the motor is used (constant torque, variable torque, etc.). **See Figure 1-10.**

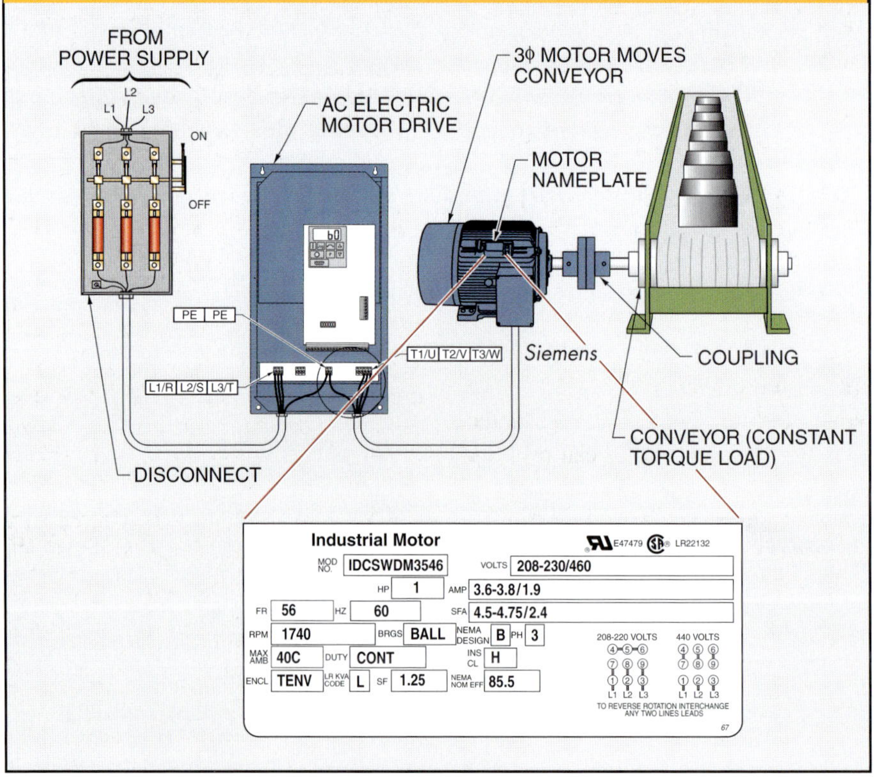

Figure 1-10. The amount of boost a motor requires is dependent upon the type of load the motor operates.

Motor Insulation Ratings

Insulation breakdown from exposure to heat is the main cause of motor failure. The National Electrical Manufacturers Association (NEMA) provides a rating for motor insulation according to its thermal endurance. The four motor insulation classes are Class A, Class B, Class F, and Class H and are listed on the motor's nameplate. Because of its low temperature rating, Class A is the least common insulation in use. Class B insulation is most commonly used on 60 Hz motors, while Class F is most commonly used on 50 Hz motors. Class H is the highest rated insulation for temperature and should be used whenever possible. **See Figure 1-11.**

MOTOR TORQUE

A motor can only produce work if its shaft turns a load. A motor must produce enough torque to start turning the load and keep it turning as required. *Torque* is the force that produces rotation. *Motor torque* is the force that produces rotation in a motor. A motor's operating torque, speed, and horsepower ratings determine the work the motor can produce.

A motor connected to a load produces four types of torque. The four types of torque are locked rotor torque (LRT), pull-up torque (PUT), breakdown torque (BDT), and full-load torque (FLT). **See Figure 1-12.**

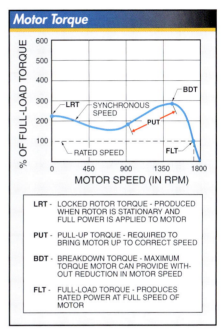

Figure 1-12. The four types of torque produced by a motor are locked rotor torque (LRT), pull-up torque (PUT), breakdown torque (BDT), and full-load torque (FLT).

Motor Insulation Operating Temperatures	
NEMA & IEC Classification	**Maximum Operating Temperatures**
A	221°F (105°C)
B	266°F (130°C)
F	311°F (155°C)
H	356°F (180°C)

Figure 1-11. Maximum operating temperatures for motor insulation are classified by the National Electrical Manufacturer's Association (NEMA) and the International Electrotechnical Commission (IEC).

Locked rotor torque (LRT) is the torque a motor produces when its rotor is stationary and full power is applied to the motor. Motors must produce enough locked rotor torque to start the shaft moving. Locked rotor torque is also referred to as breakaway or starting torque.

Pull-up torque (PUT) is the torque required to bring a load up to its rated speed. If a motor cannot produce enough pull-up torque, the motor may turn the shaft but not at rated speed. Pull-up torque is also referred to as accelerating torque.

Breakdown torque (BDT) is the maximum torque a motor can provide without an abrupt reduction in motor speed. As the load on a motor increases, the motor produces more torque until the load is too great for the motor to turn it at the motor's rated speed.

Full-load torque (FLT) is the torque required to produce the rated power at the full speed of a motor. Although full-load torque is equal to the motor's nameplate rated power at the motor's rated voltage and current, for most applications, an electric motor should not have to work 100% (full-load) to drive the load.

Motor Torque, Speed, and Horsepower Relationship

A motor's operating torque, speed, and horsepower rating determines the work the motor can produce. Operating torque, speed, and horsepower are interrelated when applied to driving a load. If a motor is fully loaded, it produces full-load torque. If a motor is underloaded, it produces less than full-load torque. If a motor is overloaded, it must produce more than full-load torque to keep the load operating at the motor's rated speed. **See Figure 1-13.**

Motor Load Types

Motors are used to drive many different types of loads. Motor loads require constant torque (CT), variable torque (VT), or constant horsepower (CH) when operating at different speeds. Understanding each type of motor load is required because motor drives must be set or programmed for the type of load the motor is to drive. **See Figure 1-14.**

Constant Torque

A *constant torque load (CT)* is a load in which the motor torque requirement remains constant. Any change in operating speed requires a change in horsepower. Constant torque loads include loads that produce friction, such as conveyors, gear driven machines, load-lifting equipment, and other loads that can operate at varying speeds.

Variable Torque

A *variable torque load (VT)* is a load that requires a varying torque and horsepower at different speeds. With a variable torque load, the motor must work harder to deliver more output at a faster speed. Both torque and horsepower are increased with increased speed. Variable torque loads include pumps, fans, mixers, and agitators.

Motor Torque, Speed, and Horsepower Characteristics

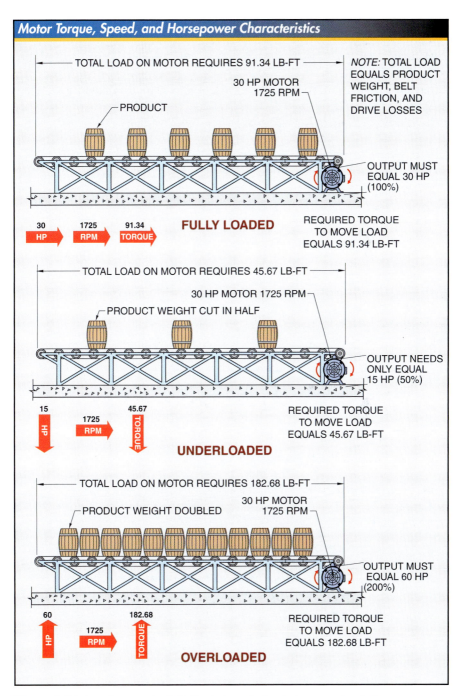

Figure 1-13. A motor may be fully loaded, underloaded, or overloaded.

Motor Load Types

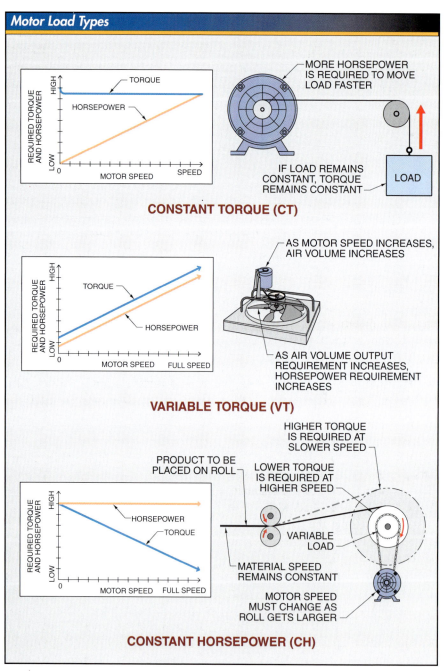

CONSTANT TORQUE (CT)

VARIABLE TORQUE (VT)

CONSTANT HORSEPOWER (CH)

Figure 1-14. Motor loads may require constant torque (CT), variable torque (VT), or constant horsepower (CH) when operating at different speeds.

Constant Power

A *constant horsepower load (CH)* is a load that requires high torque at low speeds and low torque at high speeds. Since the torque requirements decrease as speed increases, the horsepower remains constant. Constant horsepower loads include paper, metal, and fabric winding equipment used in production manufacturing facilities.

When a drive is used to control a motor, the drive is set or programmed for the load type. Setting or programming a load type is usually a matter of determining the load type and setting or programming the drive for a specific type of application.

If verification, documentation, or analysis is required, a portable oscilloscope can be used to capture and display measurements taken under different settings and circuit operating conditions. For example, separate measurements can be captured when the motor drive is set to the "CT" (constant torque) position and the "P&F" (pumps and fans) position. When in the CT setting, the drive outputs higher power at lower and higher speeds to maintain constant torque on the load. Also, when in the P&F setting, the drive outputs less power at low speeds and increases power as more liquid or air is moved at higher speeds. **See Figure 1-15.**

ELECTRIC MOTOR DRIVES

A motor drive is an electronic unit designed to control the speed of a motor using solid-state devices. Motor drives are divided into three sections and can be AC or DC drives, with AC drives as the most common. Traditionally, magnetic motor starters were used to control electric motors. However, motor drives are now the preferred method of controlling motors. Magnetic motor starters are sometimes used in basic applications to turn motors on and off and provide overload protection.

In addition to turning motors on and off and providing overload protection, motor drives are different from motor starters as they have other features such as providing speed control, timed acceleration and deceleration, motor-starting boost, fault monitoring, programmable set speeds, different stopping methods, providing pulse width modulations, and other control functions. Most parameters are programmed into the motor drive and are visible through operating displays. The length of the cable used to connect the motor drive to the motor also affects the motor's operation.

Motor Drive Features

Electric motor drives include several features that make them versatile, cost effective, and energy efficient. Motor drives have the ability to have operating parameters programmed into them and offer fault indicators that display detected drive and motor faults. Most motor drives also include built-in metering functions that measure and display basic operation parameters. The parameters are set to best match the system's intended operation. The fault indicators and meter operating displays also help determine problems when troubleshooting.

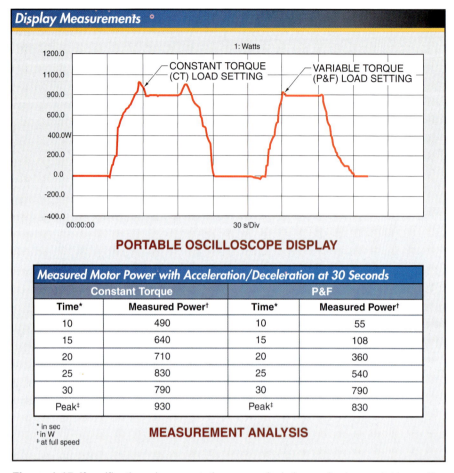

Figure 1-15. If verification, documentation, or analysis is required, a portable oscilloscope can be used to capture and display measurements taken under different settings and circuit operating conditions.

Parameter Programming. Parameter menu formatting and methods of programming vary based on manufacturer, but all motor drives offer programming of operating conditions. Most motor drives offer various parameters that can be programmed to customize the drive and motor to a specific application. However, only a few are required to be set or programmed so the drive best matches the motor and application. Although most motor drives include numerous parameters to customize a drive to a motor and application, normally most parameters do not have to be reprogrammed from the factory default settings. To simplify programming, some manufacturers group the most commonly programmed parameters together to make programming easier. **See Figure 1-16.**

Basic Programming Parameters

Basic Programming Parameters

◯ = Stop drive before changing this parameter.

No.	Parameter	Min/Max	Display/Options	Default
P031 ◯	[Motor NP Volts] — Set to the motor nameplate rated volts.	20/Drive Rated Volts	1 VAC	Based on Drive Rating
P032 ◯	[Motor NP Hertz] — Set to the motor nameplate rated frequency.	10/240 Hz	1 Hz	60 Hz
P033	[Motor OL Current] — Set to the maximum allowable motor current.	0.0/(Drive Rated Amps x 2)	0.1 Amps	Based on Drive Rating
P034	[Minimum Freq] — Sets the lowest frequency the drive will output continuously.	0.0/240.0 Hz	0.1 Hz	0.0 Hz
P035 ◯	[Maximum Freq] — Sets the highest frequency the drive will output.	0/240 Hz	1 Hz	60 Hz
P036 ◯	[Start Source] — Sets the control scheme used to start the drive.	0/5	0 = "Keypad" (1) 3 = "2-W Lvl Sens" 1 = "3-Wire" 4 = "2-W Hi Speed" 2 = "2-Wire" 5 = "Comm Port"	0

(1) When active, the Reverse key is also active unless disabled by A095 [Reverse Disable].

Figure 1-16. Basic programming parameters consist of the most commonly used programming parameters.

Operating Displays. Most motor drives monitor and display operating conditions such as output voltage, current, power, frequency, and temperature. Display parameters are used to give a visual display of operating conditions, which can be used when installing, testing, operating, and troubleshooting a circuit using a motor drive. Observing the drive's operating display also helps determine the types of additional test instruments that should be used when troubleshooting the system. **See Figure 1-17.**

Operating Displays

Display Parameters

No.	Parameter	Min/Max	Display/Options			
d001	[Output Freq]	0.0/[Maximum Freq]	0.1 Hz			
d002	[Commanded Freq]	0.0/[Maximum Freq]	0.1 Hz			
d003	[Output Current]	0.00/(Drive Amps x 2)	0.01 Amps			
d004	[Output Voltage]	0/Drive Rated Volts	1 VAC			
d005	[DC Bus Voltage]	Based on Drive Rating	1 VDC			
d006	[Drive Status]	0/1 (1 = Condition True)	Bit 3 Decelerating	Bit 2 Accelerating	Bit 1 Forward	Bit 0 Running
d007- d009	[Fault x Code]	F2/F122	F1			

Figure 1-17. Display parameters are used to give a visual display of operating conditions.

Fault Displays. Motor drives include built-in monitoring features to track system operation, automatically shut down the system if a problem is detected, and display a fault code and/or related information. As with operating displays, fault displays also help determine the types of additional test instruments that should be used when troubleshooting the system.

Main Drive Sections

The three main sections of an AC motor drive are the converter section, DC bus section, and inverter section. **See Figure 1-18.** The converter section (rectifier) receives the incoming AC voltage and changes the voltage to DC. If required, the converter section also steps up or steps down the input voltage to match the required output voltage.

The DC bus section filters the voltage and maintains the proper DC voltage level. The DC bus section delivers the DC voltage to the inverter section for conversion back to AC voltage. The inverter section changes the DC back into three-phase AC for delivery to the motor. The inverter section controls motor torque and speed.

Pulse Width Modulation

Pulse width modulation (PWM) is a method of controlling the amount of voltage sent to the motor. AC motor drives must control the amount of voltage produced in order to control the speed and torque of a motor. Pulse width modulation controls the amount of voltage output by converting the DC voltage into fixed values of individual DC pulses using the high-speed switching of transistors. By varying the width of each pulse (time on) and/or varying the frequency, the voltage can be increased or decreased. The greater the width of individual pulses, the higher the DC voltage output. **See Figure 1-19.**

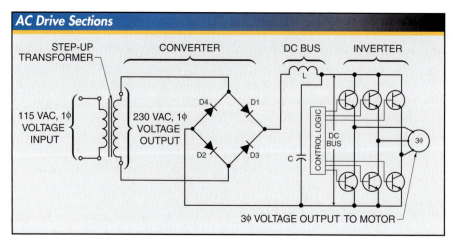

Figure 1-18. The three main sections of an AC drive are the converter, DC bus, and inverter sections.

Pulse Width Modulation

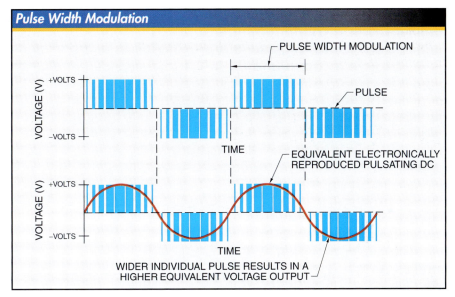

Figure 1-19. Pulse width modulation is used to produce a pulsating DC output.

A *carrier frequency* is the frequency that controls the rate at which solid-state switches in the inverter of a PWM motor drive turn on and off. The higher the carrier frequency, the more individual pulses are present to reproduce the fundamental frequency. A *fundamental frequency* is the frequency of the voltage used to control motor speed. The carrier frequency of most motor drives can range from 1 kHz to approximately 16 kHz. The higher the carrier frequency, the closer the output sine wave is to the pure fundamental frequency sine wave. However, the higher the carrier frequency, the greater the amount of transient voltage produced by switching power on/off and the higher the produced heat.

When testing or troubleshooting an installed drive, a portable oscilloscope can be used to closely analyze the drive's PWM output. First, each phase should be reviewed, and it should be verified that there are no reflections present. When the drive is set for a carrier frequency of 5 kHz, there are less individual pulses present to reproduce the fundamental frequency. When the drive is set for a carrier frequency of 16 kHz, there are more pulses present to reproduce the fundamental frequency. However, a carrier frequency of 16 kHz can have an impact on harmonics and potentially shorten the allowable cable length between the drive and motor. **See Figure 1-20.**

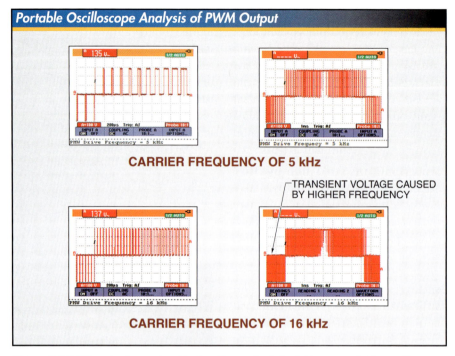

Figure 1-20. When testing or troubleshooting an installed drive, a portable oscilloscope can be used to closely analyze the drive's PWM output.

TECH TIP

Although most motors have built-in safeguards to prevent operating a motor at frequencies higher than its rated frequency, some motors can be operated well above their rated frequency, creating a hazardous situation. Never operate a motor at frequencies higher than its rated frequency.

Cable Length

In any electrical system, the distance between components and devices affects operation. The primary limit to distance between a magnetic motor starter and the motor is the voltage drop of the conductors and should not exceed 3%. The limiting factor when a motor drive is used is the amount of voltage drop and capacitance between the hot conductors delivering power to the motor and the hot conductors to ground. Longer conductors produce higher capacitance, which causes high transient voltages in the voltage delivered to the motor. Since transient voltages are reflected into the system, the transient voltages are referred to as reflective wave spikes, or reflective waves. As conductor length increases and an electric motor drive's output carrier frequency increases, the transient voltages become larger. **See Figure 1-21.**

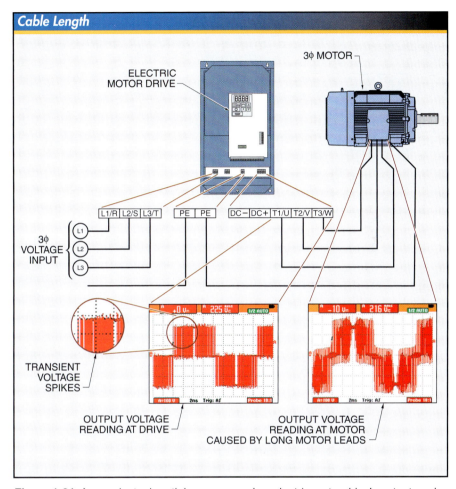

Figure 1-21. As conductor length increases and an electric motor drive's output carrier frequency increases, the transient voltages become larger.

The higher the drive's carrier frequency, the more electromagnetic noise produced and thus the shorter the cable length between the motor and motor drive should be. Less cable length reduces noise problems and is required in order to comply with certain regulations, such as the European EMC regulations.

2

Safety and Test Measurements

Troubleshooting electrical equipment requires using test instruments to take measurements on energized electrical circuits and components. Working on and near energized electrical circuits can cause electrical shocks, burns, and injuries. To reduce the chance of an electrical shock or injury, proper safety measures must be followed and personal protective equipment (PPE) must be worn and used. To locate the problem, there must be an understanding of test measurements. When using portable oscilloscopes, understanding both the numerical display and waveform display is required.

SAFETY

Safety rules and standards vary from country to country, and everyone should know and follow local regulations. However, the underlying ideas of caution and common sense are universal, and some general recommendations about safety can be made.

When installing or modifying electrical circuits, energized equipment and circuits can be turned off and locked in the OFF position. *Lockout* is the process of removing the source of electrical power and installing a lock that prevents the power from being turned on. *Tagout* is the process of placing a danger tag on the source of electrical power, which indicates that the equipment may not be operated until the danger tag is removed. A danger tag has the same importance and purpose as a lock and is only used alone when a lock does not fit onto the disconnect device.

Electrical power must be locked and tagged when electrical power is not required to be ON to perform the required service. However, electrical power must be ON when testing and troubleshooting a circuit using test instruments. To help prevent an electrical shock or any other type of injury, personal protective equipment must be worn and used, and required safety procedures and rules must be followed. Additionally, only qualified personnel must work on or near electrical equipment and circuits.

Following proper safety procedures also includes understanding which test instruments to use for specific testing procedures, understanding safe test instrument use, reading and following manufacturer procedures and recommendations, and verifying that the test instruments are rated for usage in the area, with the equipment, or on the circuit to be tested. If the original test instrument's operation manual is not available, most test instrument manufacturers provide the manuals through the Internet or will provide a replacement copy upon request.

Personal Protective Equipment

Personal protective equipment (PPE) is clothing and/or equipment worn by a person to reduce the possibility of injury in the work area. PPE must be worn anytime work or testing is performed on energized exposed electrical circuits.

PPE includes protective clothing, head protection, eye protection, ear protection, hand and foot protection, back and knee protection, and rubber insulating matting. **See Figure 2-1.**

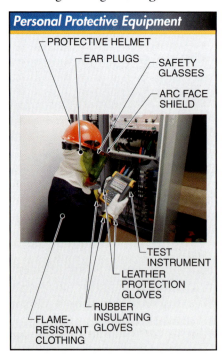

Personal Protective Equipment

PROTECTIVE HELMET

EAR PLUGS

SAFETY GLASSES

ARC FACE SHIELD

TEST INSTRUMENT

LEATHER PROTECTION GLOVES

RUBBER INSULATING GLOVES

FLAME-RESISTANT CLOTHING

Figure 2-1. Personal protective equipment (PPE) is used when taking electrical measurements to reduce the possibility of an injury.

All PPE must meet designated codes and standards, such as applicable American National Standards Institute (ANSI) standards and Occupational Safety and Health Administration (OSHA) 29 Code of Federal Regulations (CFR) 1910, Subpart I – *Personal Protective Equipment,* sections 1910.132 to 1910.138.

Qualified Persons

A *qualified person* is a person who is trained and has special knowledge of the construction and operation of electrical equipment or a specific task and is trained to recognize and avoid electrical hazards that might be present with respect to the equipment or specific task. Only a qualified person should work on or near any electrical equipment. For example, in the United States, NFPA 70E® states, "only qualified persons shall perform testing on or near live parts operating at 50 V or more." In many other countries the voltage limit is as low as 24 V. A qualified person must have the following:
- knowledge of the type of PPE that is required for specific areas and required test procedures
- proper training in PPE usage, testing procedures, required laws, rules, codes, standards, company procedures, and any other required knowledge to ensure the safest working environment possible
- knowledge of which tools and test instruments may be required and their proper and safe usage for specific test procedures

Test Instrument Measurement Categories (CAT Ratings)

The *International Electrotechnical Commission (IEC)* is an organization that develops international standards for electrical equipment. IEC standard 61010 classifies the applications in which test instruments are used into four measurement categories: CAT I, CAT II, CAT III, and CAT IV. **See Figure 2-2.** The CAT ratings determine the maximum magnitude of voltage surge (transient voltage) a test instrument can withstand when used on a power distribution system.

A *transient voltage* is an undesirable, momentary voltage pulse that varies in amplitude and energy level depending upon the source (solenoid coil or lightning strike) of the transient.

Test equipment with high CAT ratings enable technicians to work more safely and on more types of electrical applications than equipment with low CAT ratings.

In motor drive applications, there is a positive and a negative voltage relative to ground or with output voltages in delta configurations. Once a test instrument is connected, with each new connection made, one connection may be a high safety risk. This is because it is made to a higher positive or negative voltage, while the other connection may be at ground potential and be a low safety risk, such as in a wye configuration.

In other applications, each of the test connections may be at a high potential (one positive and one negative, relative to ground), which means that establishing each of these connections has certain safety risks. For example, a delta connection has safety risks because each conductor carries about the same voltage level above ground.

WARNING: The working environment for a variable speed motor drive may vary from installation to installation. Therefore, it is good practice to consult the motor drive manufacturer and qualified person for information on proper safety procedures.

There are differences between a CAT II–1000 V meter and a CAT III–600 V meter. While a CAT II–1000 V meter has a higher working voltage than a CAT III–600 V meter, a CAT III–600 V meter can safely withstand six times the current and six times the power of a CAT II–1000 V meter.

The recommendations for test instruments used on variable speed drives include the following:

• Instruments should have a minimum rating of CAT IV–600 V or CAT III–1000 V.

• Instruments that are rated CAT III–1000 V and CAT IV–600 V enable safe work with a broad range of electrical applications.

Note: For industrial motor drive applications, test instruments rated CAT I and CAT II should be avoided unless when performing work on small, fractional-horsepower drives that are plugged into wall receptacles. Avoid meters that claim to be "designed to meet" IEC 61010 specifications or that do not carry the test certification of an independent testing lab such as UL, CSA, VDE, TÜV, or MSHA. These meters do not meet all requirements associated with the specifications for which they claim to be designed.

Safety guidelines for taking test measurements on electric motor drives include the following:

Test Instrument Measurement Categories (CAT Rating)

Category	In Brief	Examples
CAT I	Electronic	• Protected electronic equipment • Equipment connected to (source) circuits in which measures are taken to limit transient overvoltages to an appropriately low level • Any high-voltage, low-energy source derived from a high-winding resistance transformer, such as the high-voltage section of a copier
CAT II	1φ receptacle-connected loads	• Appliances, portable tools, and other household and similar loads • Outlets and long branch circuits • Outlets at more than 30′ (10 m) from CAT III source • Outlets at more than 60′ (20 m) from CAT IV source
CAT III	3φ distribution, including 1φ commercial lighting	• Equipment in fixed installations, such as switchgear and polyphase motors • Bus and feeder in industrial plants • Feeders and short branch circuits and distribution panel devices • Lighting systems in larger buildings • Appliance outlets with short connections to service entrance
CAT IV	3φ at utility connection, any outdoors conductors	• Refers to the origin of installation, where low-voltage connection is made to utility power • Electric meters and primary overcurrent protection equipment • Outside and service entrance, service drop from pole to building, run between meter and panel • Overhead line to detached building

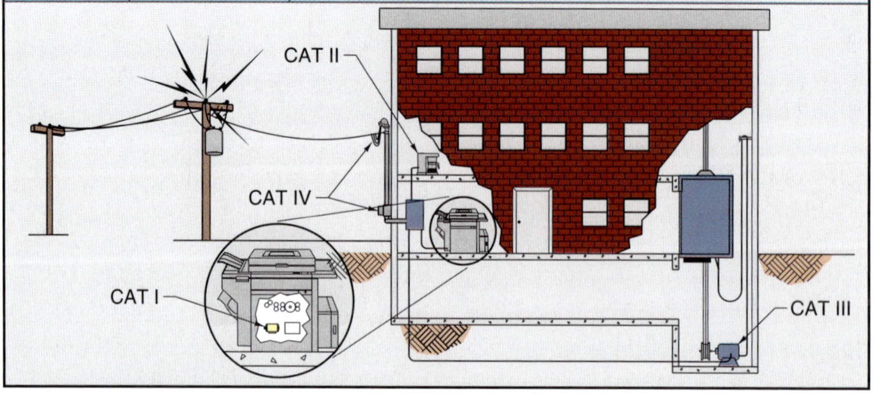

Figure 2-2. The International Electrotechnical Commission (IEC) standard 61010 classifies the applications in which test instruments are used into four measurement categories (Category I–Category IV).

- Work on de-energized circuits whenever possible, using proper lockout/tagout procedures. Verify that others know you are working on the system.
- Use proper PPE and hand tools when working on live circuits.
- Do not measure voltages above the working voltage rating indicated on the test instrument.
- Do not measure in high-energy environments for which the instrument is not rated.
- Connect the ground lead first, followed by the hot lead.
- Keep one hand in pocket to reduce the possibility of a closed circuit across the chest and through the heart.
- If possible, hang meter by strap or prop up to avoid holding the meter.
- Use the three-step test method, which involves testing a known live circuit, testing the target circuit, and retesting the live circuit to verify that the meter is still functioning properly.

To avoid electrical shock or other injury when taking measurements on CAT III circuits, voltage divider probes (for example, 10:1, 100:1, 1000:1, etc.) that are only rated IEC 61010 CAT I or CAT II should not be used. It should be verified that the scope probe used is rated for the measurement category environment where work is being performed. The combination of CAT III–rated test instruments and CAT I– or CAT II–rated accessories should be avoided.

Whenever possible, a multichannel test instrument should be used and all test connections made using one common reference point. In three-phase power systems, this is typically the "neutral" or common return line. In other systems, it can be the common reference or ground wire. By using a multichannel test instrument with one common reference point, the voltages applied to the test instrument and to all associated wiring are minimized, reducing safety risks. Also, the actual voltage levels that the instrument has to withstand are lower, which also reduces safety risks.

Note: When using multichannel test instruments, it may be necessary to make a test connection to the three-phase system, where one phase connection is used as the common reference for two test instrument channels. The phase-to-phase voltages can then be measured. The "common" is at a high voltage and should be treated with the same safety considerations as the test inputs.

TEST MEASUREMENTS

Most test instruments are designed to take several different types of specific measurements. Understanding the type of measurements that can be taken and analysis of their subsequent measurements is a requirement for understanding how a circuit, system, component, or piece of equipment is operating. An understanding of the types of voltage and current present in a circuit, as well as their measured values, is also required when performing maintenance and troubleshooting tasks.

Measurements taken with a portable oscilloscope are typically displayed as a number or waveform. **See Figure 2-3.**

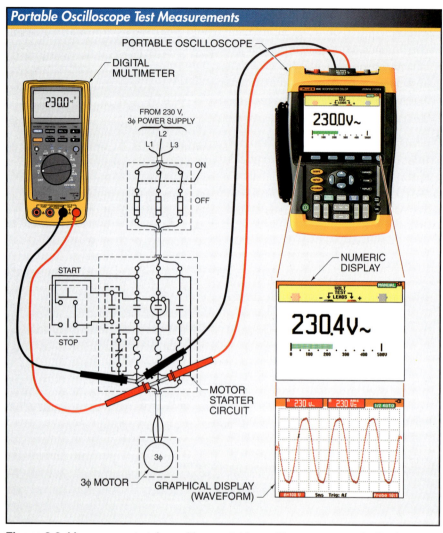

Figure 2-3. Measurements taken with a portable oscilloscope are typically displayed as a number or waveform.

In some applications, numerical values provide enough information to solve the problem. However, when troubleshooting more complex problems, a visual sine wave of the measured value (voltage, current, power, or resistance) is more valuable to the troubleshooter.

Multichannel portable oscilloscopes can take multiple measurements simultaneously and measure multiple aspects of each signal. This enables the technician to have a faster and more complete understanding of circuit behavior.

Voltage and Current Types

Voltage is the amount of electrical "pressure" in a circuit. Voltage is measured in volts (V) and can be either direct (DC) or alternating (AC). *Current* is the flow of electrons in an electrical circuit. Current is measured in amperes (A). As with voltage, current is either direct or alternating.

DC Circuits. DC power sources have a positive and negative terminal. The positive and negative terminals establish polarity in a circuit. *Polarity* is the positive (+) or negative (–) electrical state of an object. All points in a DC circuit have polarity. DC can be pure DC, half-wave DC, full-wave DC, filtered DC, or a varying DC depending upon the source of the DC. For example, a DC power source such as a battery can produce pure DC. DC that is rectified (changed) from AC to DC can be half-wave rectified AC or full-wave rectified AC and can be filtered using capacitors to make the DC more pure. **See Figure 2-4.**

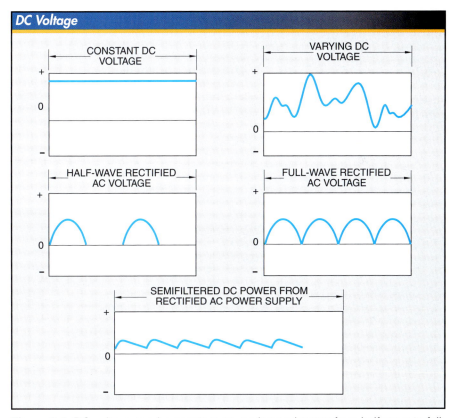

Figure 2-4. DC voltage can be constant or varying and range from half-wave to full-wave rectified AC.

AC Circuits. Unlike the steady DC voltage produced by batteries, the voltage produced by electric generators alternates back and forth between positive and negative values as the generator rotor revolves through 360 degrees. A voltage that alternates between positive and negative values produces a similarly alternating current (AC), from which AC power gets its name.

When viewed on a portable oscilloscope, an AC voltage can be seen to vary in a smoothly varying sine wave over each 360° cycle. **See Figure 2-5.** Starting from 0 V at 0°, the sine wave reaches its peak positive value at 90°, returns to 0 V at 180°, increases again to its peak negative value at 270°, and returns to 0 V at 360°. An alternation is half of a cycle (180°). There is one positive alternation and one negative alternation in each cycle of an AC sine wave.

AC Voltage. AC voltage values can be measured and displayed as an average, root-mean-square (rms), peak, or peak-to-peak value. The *average voltage (V_{avg})* of a sine wave is the mathematical mean of all instantaneous voltages of a half cycle of a sine wave. The average value is equal to 0.637 times the peak value of a pure sine wave. The *root-mean-square (effective) voltage (V_{rms})* is the voltage that produces the same amount of heat in a pure resistive circuit as is produced by an equal DC voltage. The rms value is equal to 0.707 times the peak value in a sine wave. The *peak voltage (V_{max})* is the maximum value of either the positive or negative alternation. The positive and negative alternations are equal in a sine wave. The *peak-to-peak voltage (V_{p-p})* is the value measured from the maximum positive alternation to the maximum negative alternation. **See Figure 2-6.**

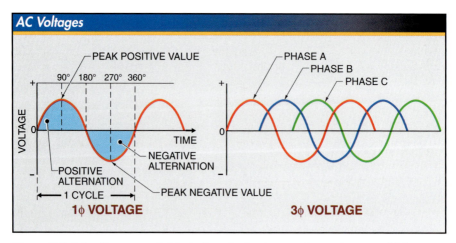

Figure 2-5. AC voltages are distributed as single-phase or three-phase, and AC motors are rated for use on single-phase or three-phase systems.

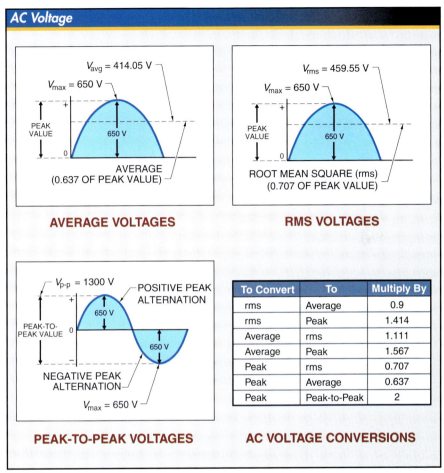

AC Voltage

AVERAGE VOLTAGES

$V_{avg} = 414.05$ V
$V_{max} = 650$ V

PEAK VALUE

650 V

AVERAGE
(0.637 OF PEAK VALUE)

RMS VOLTAGES

$V_{rms} = 459.55$ V
$V_{max} = 650$ V

PEAK VALUE

650 V

ROOT MEAN SQUARE (rms)
(0.707 OF PEAK VALUE)

PEAK-TO-PEAK VOLTAGES

$V_{p-p} = 1300$ V
POSITIVE PEAK ALTERNATION

650 V

PEAK-TO-PEAK VALUE

650 V

NEGATIVE PEAK ALTERNATION

$V_{max} = 650$ V

AC VOLTAGE CONVERSIONS

To Convert	To	Multiply By
rms	Average	0.9
rms	Peak	1.414
Average	rms	1.111
Average	Peak	1.567
Peak	rms	0.707
Peak	Average	0.637
Peak	Peak-to-Peak	2

Figure 2-6. AC voltage can be measured and displayed as an average, rms, peak, or peak-to-peak value in the form of sine waves.

Portable Oscilloscope Displays

A portable oscilloscope displays the measured electrical values as a number or waveform. A portable oscilloscope display screen contains scribed horizontal and vertical axes. The horizontal (x) axis represents time. The vertical (y) axis represents the amplitude of the waveform. A grid on the display screen divides the display into segments that can be used to measure the voltage and frequency of displayed waveforms.

A *trace* is a line that sweeps across the display screen of a portable oscilloscope to display a signal's amplitude over time. The trace sweep is adjusted with the horizontal time base setting. **See Figure 2-7.**

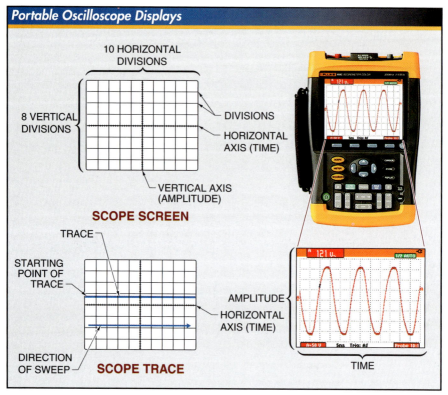

Figure 2-7. Horizontal and vertical axes divide an oscilloscope display screen into equal divisions while a trace is established on the screen before a circuit under test is connected.

Multichannel portable oscilloscopes can display up to four traces at a time. This ability enables the technician to simultaneously view and compare multiple signals and different aspects of each signal such as voltage and current.

Horizontal and Vertical Positioning. In order to better understand displayed measurements on their screens, portable oscilloscopes have mechanisms that allow adjustments to be made to the displayed information. Typical adjustments can be made for horizontal and vertical positioning, time per horizontal division, volts per vertical division, and movement of cursor positions. **See Figure 2-8.**

Multichannel portable oscilloscopes can display signals for up to four different channels.

Horizontal and Vertical Positioning Adjustments

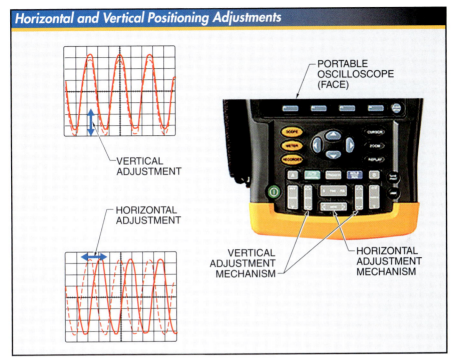

Figure 2-8. Portable oscilloscopes have mechanisms that allow horizontal and vertical adjustments to be made to the displayed information.

The horizontal and vertical positioning adjustment can be used to overlay two different measured values. Overlaid measurements can indicate if the two measurements are in-phase or out-of-phase with each other. The more voltage and current are out-of-phase, the greater the difference between the circuit's true power and apparent power, and therefore the worse the circuit's power factor (PF). **See Figure 2-9.**

Being able to visualize and measure phase shifts help identify a circuit's true power, apparent power, and power factor. For example, if the measured value of an inductive load indicates

that the circuit has a poor power factor of 32%, it also indicates that the power supply (transformer) has to deliver 4.2 kVA in order to produce 1.3 kW of power output by the load. An inductive load, such as motor windings or solenoids, causes the phase shift between voltage and current. In an inductive load, current lags voltage.

TECH TIP

Oscilloscope controls that can be adjusted include intensity, focus, horizontal positioning, volts per division, and time per division.

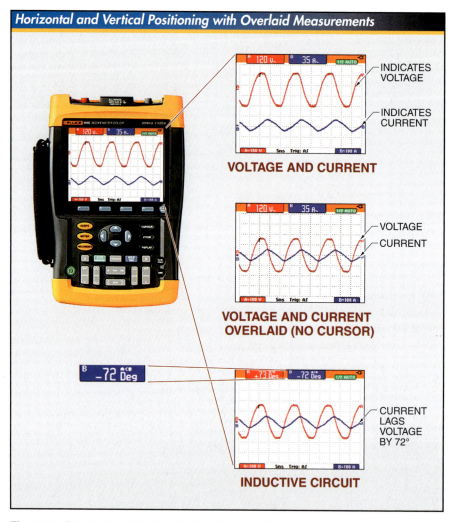

Figure 2-9. The horizontal and vertical positioning adjustment on a portable oscilloscope can be used to overlay two different measured values.

Waveforms

Circuit measurements are shown on the display screen of a portable oscilloscope as waveforms in different colors that vary in size and shape. Understanding the meaning of the waveform size, shape, and color is essential when analyzing or troubleshooting a circuit. Waveforms can represent linear or nonlinear loads. **See Figure 2-10.**

Linear Loads. A *linear load* is a load in which current increases proportionately as the voltage increases, and current decreases proportionately as voltage decreases. Linear loads, such as heating elements, do not distort the voltage or current sine wave regardless of whether the voltage and current are in-phase or out-of-phase.

Waveforms

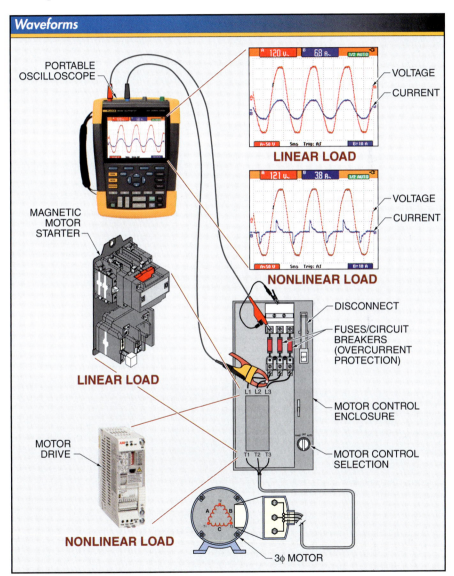

Figure 2-10. Depending on the circuit or equipment measured, waveforms as shown on a portable oscilloscope display can represent linear or nonlinear loads.

Nonlinear Loads. A *nonlinear load* is a load in which the instantaneous load current is not proportional to the instantaneous voltage. Nonlinear loads, such as computers, printers, and motor drives, distort the waveform representing current.

Waveform Analysis. A displayed waveform can be analyzed to indicate the quality of circuit operation and identify problems. While displayed waveforms can indicate proper operation, they can also identify such problems as loss of signal, distorted signals, transient voltage, or ringing. **See Figure 2-11.**

Waveform Analysis

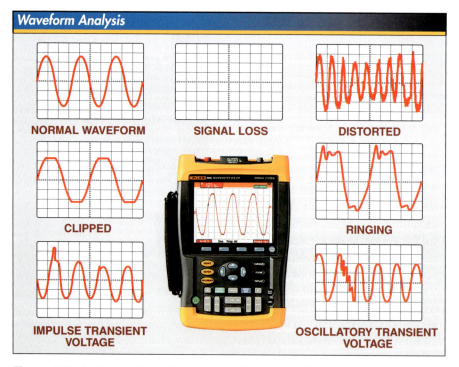

NORMAL WAVEFORM

SIGNAL LOSS

DISTORTED

CLIPPED

RINGING

IMPULSE TRANSIENT VOLTAGE

OSCILLATORY TRANSIENT VOLTAGE

Figure 2-11. A displayed waveform can identify such problems as loss of signal, distorted signals, clipped signals, transient voltages, or ringing.

Troubleshooting is the systematic elimination of various sections of a system to locate a malfunctioning device or component. Troubleshooting can include testing mechanical systems, hydraulic systems, pneumatic systems, electronic/communication systems, electrical systems, or other systems. Troubleshooting often requires inspecting and testing several different systems, since most systems are interconnected and operate together. Troubleshooting several systems applies when troubleshooting motors and motor drives since motors are usually used in most systems and applications.

When troubleshooting electric motors, motor drives, and electrical systems, the tests to be performed are determined based on the required information needed to help determine operating conditions and/or problems. The required test instruments are selected based on the test to be performed and required documentation.

TROUBLESHOOTING TEST INSTRUMENTS

When testing electrical systems and components, electrical test instruments are used. Some electrical test instruments are designed for one specific purpose while others are designed for a range of tasks. Every electrical test instrument has advantages and disadvantages that determine its usefulness in a certain situation. Advantages and disadvantages include such factors as cost, number and types of individual measurements available on the meter, ease in understanding functions and usage, ability to record and document measurements, CAT rating, and accuracy of measurements. For most troubleshooting tasks, the more complex the problem, the more complex the test instrument requirements. The most common test instruments used are digital multimeters (DMMs), power quality meters, portable oscilloscopes, insulation testers, and noncontact test instruments such as digital thermometers or thermal imagers.

General-Purpose DMMs

A *digital multimeter (DMM)* is an electrical testing instrument that measures two or more electrical properties and displays the measured properties as numerical values. A general-purpose DMM (sometimes referred to as a "tester") is typically used for basic voltage, resistance, and current measurements when testing basic devices (fuses, switches, etc.) or taking basic troubleshooting measurements (voltage or current to motor, etc.). Most general-purpose DMMs also include the ability to measure resistance and check continuity.

General-purpose DMMs can be used for taking basic voltage, resistance, and current measurements in motor control circuits that use magnetic motor starters. However, they may not be accurate on measurements taken from the output of a motor drive to the motor because the drive's carrier frequency (2 kHz to about 16 kHz) interferes with the meter's reading. It is the carrier frequency/pulse width modulation (PMW) frequency that produces the fundamental frequency (0 to 60 Hz) that the motor uses to produce rotation. Without specifically designed low-pass filters within the meter, the meter reading is distorted because the meter is trying to read the full bandwidth of both the carrier and fundamental signal.

TECH TIP

Although general-purpose DMMs work well for most applications, it is a good practice to always have an advanced meter with several attachments also available when troubleshooting.

General-purpose DMMs are used for measuring voltage at the source (disconnect, circuit panel, receptacles, etc.) and testing fuses. They can be used to measure voltage between phases, such as phase-to-neutral and phase-to-ground. When measuring between phases, verify that the voltage and current are balanced. Voltage unbalances of more than 2% or current unbalances of 12% indicate a power problem, and additional system test measurements should be taken. **See Figure 3-1.**

Standard DMMs

A standard DMM includes the basic measuring features included with a general-purpose DMM and also includes several additional features that make it useful for taking measurements and troubleshooting most circuits and components. Standard DMMs often include the ability to test diodes and capacitors directly and can also be used with add-on attachments that measure temperature, current, and pressure. **See Figure 3-2.**

Standard DMMs usually include extra selector switch positions for taking measurements such as low-impedance current and voltage measurements. They also include extra function buttons for recording or comparing measurements such as MIN/MAX (minimum and maximum), REL (relative), and PEAK or a HOLD button to freeze a measurement. Standard DMMs also include a bar graph that can be used for viewing fast changing signals in addition to the numerical displays.

Advanced DMMs

An advanced DMM includes the measuring features of general-purpose and standard DMMs and also includes several additional features. Advanced DMM models typically have five-digit displays and/or have a higher count capacity for more accurate measurements, can capture and store measurements for later viewing, can be connected to a PC for downloading measurements, and/or include a trending feature that allows measurements taken over time to be displayed as a single line on a graph.

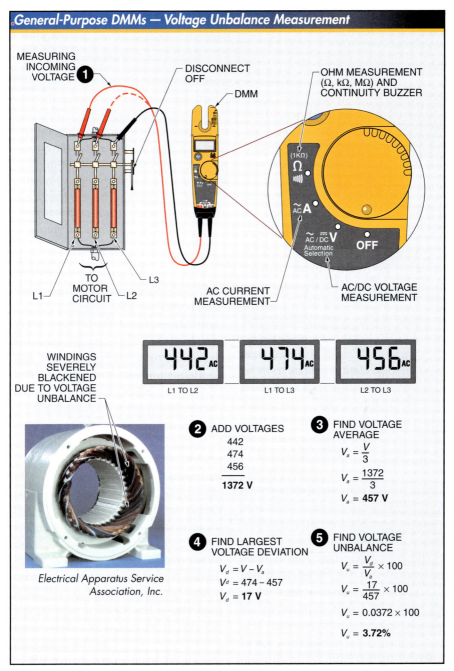

Figure 3-1. A general-purpose DMM can be used for taking basic measurements such as measuring voltage at a disconnect switch.

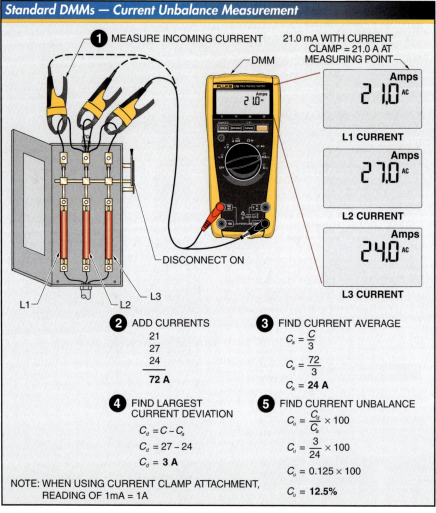

Standard DMMs — Current Unbalance Measurement

1 MEASURE INCOMING CURRENT

DMM

21.0 mA WITH CURRENT CLAMP = 21.0 A AT MEASURING POINT

Amps
$2 1.0$ AC

L1 CURRENT

Amps
$2 7.0$ AC

L2 CURRENT

Amps
$2 4.0$ AC

L3 CURRENT

DISCONNECT ON

L1 L2 L3

2 ADD CURRENTS

21
27
24
―――
72 A

3 FIND CURRENT AVERAGE

$C_a = \dfrac{C}{3}$

$C_a = \dfrac{72}{3}$

$C_a = $ **24 A**

4 FIND LARGEST CURRENT DEVIATION

$C_d = C - C_a$

$C_d = 27 - 24$

$C_d = $ **3 A**

5 FIND CURRENT UNBALANCE

$C_u = \dfrac{C_d}{C_a} \times 100$

$C_u = \dfrac{3}{24} \times 100$

$C_u = 0.125 \times 100$

$C_u = $ **12.5%**

NOTE: WHEN USING CURRENT CLAMP ATTACHMENT, READING OF 1mA = 1A

Figure 3-2. A standard DMM includes several measuring functions and features and can be used with add-on attachments.

Some of the most important features an advanced DMM offers when troubleshooting motor drives and motors include a setting specifically designed for taking voltage/frequency measurements that does not allow the carrier frequency to interfere with the measurement and a peak capture measuring mode for measuring voltage and current peak values. The peak capture mode can be used to determine if the voltage or current waveform is distorted by comparing the reading to the rms value. The peak capture mode also can be used to capture individual transient voltages or currents. **See Figure 3-3.**

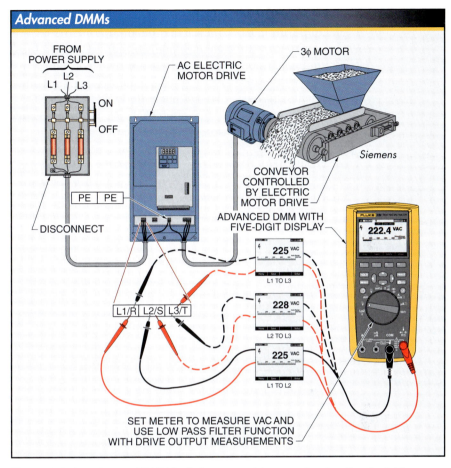

Advanced DMMs

FROM POWER SUPPLY

L1 L2 L3

ON

OFF

DISCONNECT

PE PE

L1/R L2/S L3/T

AC ELECTRIC MOTOR DRIVE

3φ MOTOR

Siemens

CONVEYOR CONTROLLED BY ELECTRIC MOTOR DRIVE

ADVANCED DMM WITH FIVE-DIGIT DISPLAY

222.4 VAC

225 VAC — L1 TO L3

228 VAC — L2 TO L3

225 VAC — L1 TO L2

SET METER TO MEASURE VAC AND USE LOW PASS FILTER FUNCTION WITH DRIVE OUTPUT MEASUREMENTS

Figure 3-3. An advanced DMM includes a special measuring function for taking measurements when a drive controls the motor so that the voltage is not interfered with by the drive's carrier frequency.

Power Quality Meters

A *power quality meter* is a test instrument that measures, displays, and records voltage, current, and power in addition to special power problems such as sags, swells, transients, and harmonics. The advantage of power quality meters is that they can measure, display, and record measurements such as voltage and current so any distortion can be viewed for a better understanding of what the problems are. Power quality meters are available in both single-phase and three-phase models. Single-phase models can be used to test three-phase circuits by connecting the meter test leads to different phases within the circuit. **See Figure 3-4.**

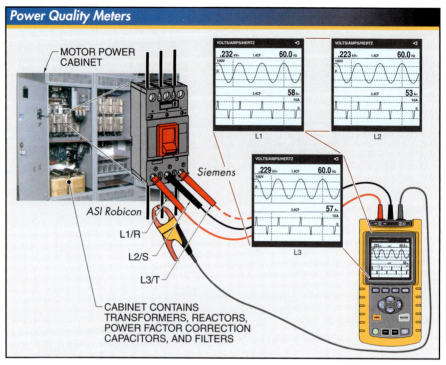

Figure 3-4. A power quality meter displays both numerical and waveform data to help determine power quality problems.

Portable Oscilloscopes

A portable oscilloscope is a test instrument that measures and displays the waveforms of high-voltage power, low-voltage control, and digital signals. Portable oscilloscopes provide the most accurate measurement and display of electrical and electronic waveforms of any type of test instrument. Portable oscilloscopes have an advantage over benchtop oscilloscopes because they are portable, handheld, battery-operated, and CAT rated for taking measurements in commercial and industrial environments. Portable oscilloscopes are ideal for testing and troubleshooting motors and motor drives when performing installation and preventive maintenance tasks. In addition to identifying the same problems as DMMs and power quality meters, they are used for more complex troubleshooting tasks such as analyzing circuit and component diagnostics, which can predict problems before they cause damage.

Portable oscilloscopes are actually two meters in one housing. In addition to a scope mode, they have a meter-measuring mode that performs the same functions as a DMM. The meter measures and displays numerical values in ohms, voltage and current, tests diodes, tests continuity, and accepts attachments for measuring temperature and pressure. **See Figure 3-5.**

Portable Oscilloscopes

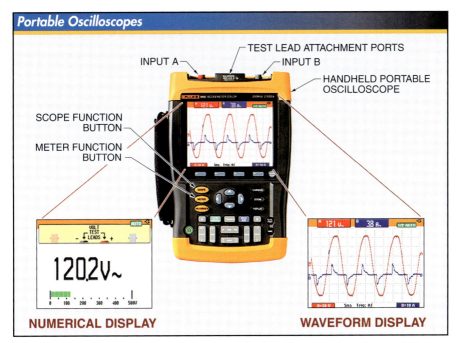

TEST LEAD ATTACHMENT PORTS

INPUT A

INPUT B

HANDHELD PORTABLE OSCILLOSCOPE

SCOPE FUNCTION BUTTON

METER FUNCTION BUTTON

NUMERICAL DISPLAY

WAVEFORM DISPLAY

Figure 3-5. Portable oscilloscopes are a combination of a meter that takes and displays digital meter readings and a scope that displays, records, and captures waveform data measurements that can be downloaded to a PC.

An advantage of portable oscilloscopes is that they display numerical values along with sine wave measurements such as voltage, current, and power over time. Another advantage is that glitches, spikes, and other waveform information that is not visible on a DMM is visible on a portable oscilloscope.

For most measurements, the vertical (y) axis represents the measured value, and the horizontal (x) axis represents time. The advantage of seeing waveform information is that the technician can pinpoint problems and verify that any circuit improvement modifications previously implemented were successful. Another advantage is that portable oscilloscopes can time-stamp measurements

such as a captured/recorded high voltage spike so the problem can be traced to the exact point in time when it occurred.

Portable oscilloscopes also have a considerably higher measurement bandwidth and sampling rate than power quality meters, which is ideal for locating electrical, electronic, and digital problems in complex circuits. *Bandwidth* is the range of frequencies that a portable oscilloscope or meter can accurately measure. *Sampling* is the process of converting a portion of an input signal into a number of discrete electrical values for the purpose of storage, processing, and display. The higher the sample rate, the more accurate the trace of the signal under test.

Portable oscilloscopes are ideal for troubleshooting motor drives and power problems because a dual-trace portable oscilloscope can measure and display two signals simultaneously. The individual traces or scope measurements can include AC, DC, and AC+DC voltages and currents, peak voltage and current, power (PF, W, VA, and VAR), frequency, PWM, and other specialized measuring functions. **See Figure 3-6.**

Multichannel portable oscilloscopes are ideal for troubleshooting motor drives and power quality problems because they can measure and display up to four signal traces simultaneously. A four-channel portable oscilloscope can compare and contrast multiple current and voltage levels and verify the integrity of waveforms by identifying transients and reflections. A problem on one phase is an indication that the problem is with a conductor, such as a bad connection. If the problem is present on all phases, the problem is with the entire cable.

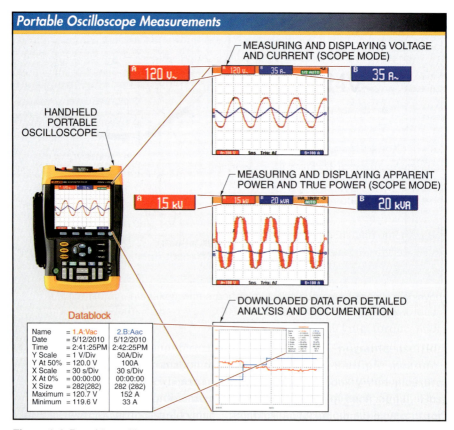

Figure 3-6. Portable oscilloscopes measure and display two signals simultaneously. Multichannel portable oscilloscopes measure and display up to four signals simultaneously.

Insulation Testers

Electrical conductors and motor windings are usually covered with some type of plasticized insulation material. The insulation material prevents current from flowing outside the designated flow path through the conductor. Eventually, insulation deteriorates over time and the resistance properties of the insulation decrease. Moisture, extreme temperatures, dust, dirt, oil, vibration, pollution, and mechanical stress or damage cause deterioration. Electric motor windings are subject to the main causes of insulation deterioration.

Although DMMs can measure resistance, they cannot accurately indicate the insulation's operating condition because they apply a low voltage from the battery of the meter to take the measurement. For an accurate measurement of insulation resistance properties, an insulation tester is used because it can measure resistance at the voltage level at which the motor, device, or component is designed to operate.

An *insulation resistance tester* is a test instrument that has specific functions for insulation resistance testing. Insulation resistance testers are typically rated to operate on a range from 0.1 MΩ to 600 MΩ. They take voltage, current, resistance, and continuity measurements from 50 V to 1000 V and read resistances from a few ohms to 2 GΩ or higher. Insulation testers are used for preventive maintenance testing because they can provide information on how well the insulation is performing over time and indicate a potential failure before it occurs. **See Figure 3-7.**

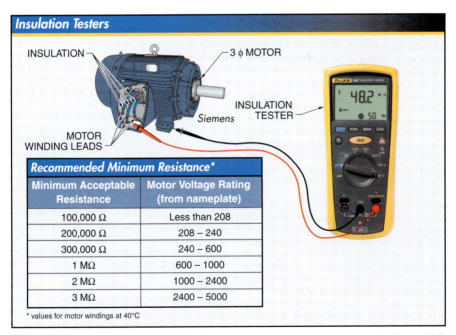

Insulation Testers

INSULATION

3 φ MOTOR

INSULATION TESTER

Siemens

MOTOR WINDING LEADS

Recommended Minimum Resistance*	
Minimum Acceptable Resistance	Motor Voltage Rating (from nameplate)
100,000 Ω	Less than 208
200,000 Ω	208 – 240
300,000 Ω	240 – 600
1 MΩ	600 – 1000
2 MΩ	1000 – 2400
3 MΩ	2400 – 5000

* values for motor windings at 40°C

Figure 3-7. An insulation tester can measure resistance at the voltage level at which a motor or component is designed to operate.

TECH TIP

While noncontact meters do not require physical contact with the area under test, proper PPE for the area where the measurement is taken must still be worn.

Noncontact Test Instruments

A *digital thermometer* is a device used to take temperature measurements on energized circuits or on moving parts without contacting the point of measurement and provide measurement readings on a digital display. Noncontact digital thermometers are used to take basic surface temperature measurements of a small area at one time without contacting the point of measurement. They can be used when troubleshooting to help determine problems caused by high temperatures. **See Figure 3-8.**

A *thermal imager* is a device that detects heat patterns in the infrared wavelength spectrum without making direct contact with the workpiece. All materials emit infrared energy in proportion to their temperature. Thermal imagers are used to locate electrical problems that cause heat, such as poor electrical connections, undersized conductors, overloaded devices, poor ventilation, and excessive moisture.

Electrical and electronic devices must not be exposed to temperatures higher than their temperature rating. High temperatures can destroy electrical insulation, devices, and components, but may not always be noticed as the cause of the problem. After the heat has caused damage, the equipment may no longer be operable so the source of the heat is no longer present.

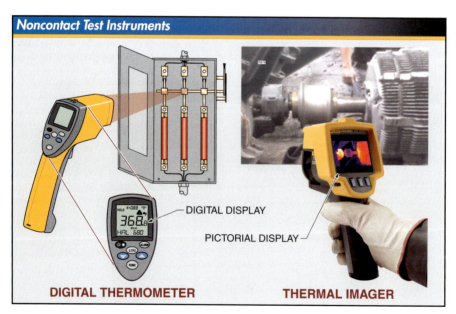

Noncontact Test Instruments

DIGITAL DISPLAY

PICTORIAL DISPLAY

DIGITAL THERMOMETER **THERMAL IMAGER**

Figure 3-8. Test instruments that can be used to test electrical components and circuits without making physical contact include digital thermometers and thermal imagers.

TROUBLESHOOTING STRATEGIES

To locate and correct a malfunction or problem quickly, troubleshooting is performed at different levels using different test instruments. The different levels are the system, equipment/unit, board/module, and component level. **See Figure 3-9.** The troubleshooting level determines the required PPE, the test instrument CAT rating, the best type of test instrument to use, the anticipated problems, and the desired documentation.

Most troubleshooting tasks begin at the equipment or unit level where the problem is noticed. Then troubleshooting is performed at the board/module and component levels where most problems are located. However, a standard troubleshooting strategy is to test each level back to the system level to identify any secondary problems that may be present within the system.

Troubleshooting Procedures

A *troubleshooting procedure* is a logical step-by-step process used to identify a malfunction or problem in a system or process as quickly and easily as possible. Using the proper test instruments helps locate and document the main problem faster and makes it easier to identify any secondary problems. To troubleshoot a system, equipment, or component apply the following procedure:

1. Obtain information by gathering technical records from the original equipment manufacturer (OEM), suppliers, contractors, operators, and maintenance departments.

2. Select the required PPE for the anticipated working environment and test instruments for the expected measurements that are to be taken.

3. Isolate the main area or section of the system or equipment that is causing the malfunction based on past experience and information obtained in step 1.

4. Test for malfunction by taking readings on the component that may be the cause of the problem or the area where the problem seems to be located. Take multiple measurements to verify proper operation or that a problem exists.

5. Once the problem is located, repair the malfunction and verify that the equipment or component is working correctly before taking the next step. *Note*: Repair may include immediate technical service, part replacement, or redeeming the OEM's or contractor's warranty.

6. After the repair is completed, take additional measurements to locate possible secondary problems.

7. Document the original problem, tests performed, and corrective action taken. List secondary problems that may need attention and possible suggestions for corrective action.

8. Prepare for future troubleshooting tasks by rechecking and servicing PPE and test instruments. Take inventory of tools and note additional tools, test instruments, or training that could have helped make the troubleshooting task easier.

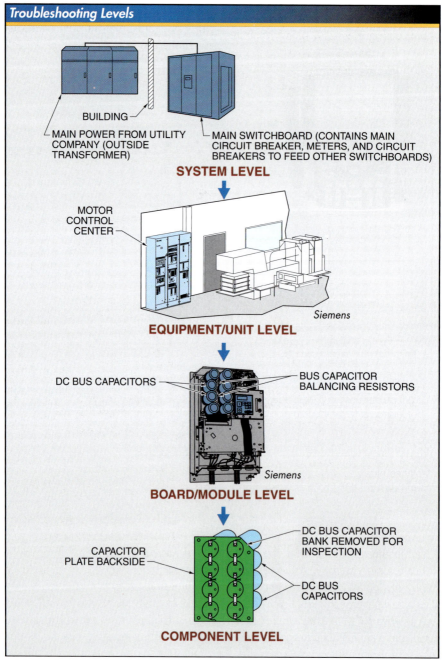

Figure 3-9. The different levels of troubleshooting electrical systems are the system, equipment/unit, board/module, and component level.

ELECTRICAL TESTS

When taking electrical measurements to help determine how a system is operating or when locating problems, the specific tests and test location must be determined. Testing or troubleshooting electrical equipment or systems begins with basic tests and progresses to more advanced tests and diagnostic troubleshooting as required.

Portable oscilloscopes can be used to take basic electrical tests and more advanced diagnostic tests.

Basic Tests

Basic electrical tests begin with taking measurements at logical and easily accessible measuring points. The most common basic tests are voltage and current tests. When performing basic tests, closely observe the equipment and surrounding area. Water, moisture, dirt, dust, and the presence of animals can cause blown fuses, tripped circuit breakers, burnt components or insulation, insufficient airflow over electrical

components, short circuits, erratic operation, higher than normal operating temperatures, or increased energy cost.

TECH TIP

No current flows in any part of a circuit when a fuse blows or a circuit breaker opens. Check the fuses or breakers first if a circuit has no power.

Voltage Tests. Electrical systems and components require a source of power that is within the proper voltage rating of the components within the system. Testing and troubleshooting start with measuring the main voltage supplied to the equipment. The voltage should be measured when the equipment is OFF and when the equipment is ON. **See Figure 3-10.**

The voltage delivered to the equipment or the system should be within −10% to +5% of the equipment nameplate rating when measured when the equipment is OFF, unless specified by the OEM. The operating voltage when the equipment is ON should not change by more than ±3% (MAX) of the voltage measured when the equipment is OFF.

In a properly sized power distribution system, the voltage measured when the equipment is OFF or ON should be the same. The larger the voltage drop between when the equipment is OFF and ON, the more likely it is that the power distribution system is overloaded or the conductors delivering the power are too long or undersized. Higher voltage drops require system testing for power issues.

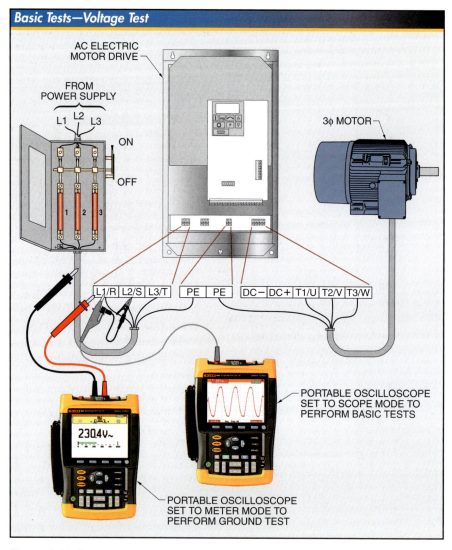

Basic Tests—Voltage Test

AC ELECTRIC MOTOR DRIVE

FROM POWER SUPPLY

L1 L2 L3

ON

OFF

3φ MOTOR

L1/R L2/S L3/T | PE PE | DC− DC+ T1/U T2/V T3/W

PORTABLE OSCILLOSCOPE SET TO SCOPE MODE TO PERFORM BASIC TESTS

PORTABLE OSCILLOSCOPE SET TO METER MODE TO PERFORM GROUND TEST

Figure 3-10. Basic electrical tests include voltage tests and start with measuring the main voltage supplied to the equipment.

When taking basic voltage tests verify that voltage measurements are taken between each phase-to-hot conductor, phase-to-neutral (ungrounded conductor), and phase-to-ground (grounded conductor). Phase-to-phase voltage measurements should not deviate more than 1% to 3%. Measurements of the voltage between various exposed metallic parts and ground should be taken to ensure correct grounding.

If a neutral conductor is included within the system, phase-to-neutral measurements verify that the neutral conductor is present for single-phase power and also indicates the type of distribution system. For example, a 208 V phase-to-phase measurement and a 120 V phase-to-neutral measurement indicates a 120/208 V, 3φ, 4-wire wye system. A 230/240 V phase-to-phase measurement and a 115/120 V phase-to-neutral measurement between phase A or phase C and 175/190 V measurement between phase B and the neutral conductor indicate a 120/240 V, 3φ, 4-wire delta system.

When taking voltage measurements where fuses are also present, verify that the fuses are in proper operating condition. Measure the voltage into the fuse (power input side) and out of the fuse (power output side) by moving the same meter test lead from the input to the output side of the fuse. If the voltage into the fuse equals the voltage out of the fuse, and the voltage across the fuse is 0 V, the fuse is good (closed). If there is no voltage out of the fuse, the fuse is bad (open).

The same basic voltage measurement test taken at the main voltage into the equipment is also taken at individual loads (motors, etc.) within the equipment. For example, the voltage delivered to a motor should be within –10% to +5% of the motor's nameplate rating when measured when the equipment is ON.

Current Tests. Basic voltage measurements are best for testing if the voltage is present and at the correct level.

However, voltage measurements do not indicate the amount of work the load is actually producing. For example, a motor that is new, inoperable, delivering half power, or delivering full power or more can have the same voltage measurement at the motor as the load.

For a more accurate indication of how loaded a motor is, a current measurement must be taken. The listed current rating on the motor nameplate is the amperage the motor or equipment should be operating at when at full power. Most motors should draw less than the nameplate rated current. **See Figure 3-11.**

Current measurement should be taken over time because unlike voltage, which usually remains the same, current changes as the required power and load changes. In three-phase systems, current readings are also a better indicator of a loss of phase than a voltage measurement.

Advanced Tests

Basic voltage and current measurements can determine certain problems such as open fuses, loss of power, overloaded motors (current readings higher than nameplate rating), and an open ground or no ground. For more complex problems, and a better understanding of how the motor or system is operating over time, advanced tests must be performed.

Advanced test measurements using DMMs or portable oscilloscopes include taking minimum and maximum, relative, peak, and temperature measurements. Advanced test measurements using a power quality meter or portable oscilloscope include taking power measurements and checking for problems with any waveform distortion.

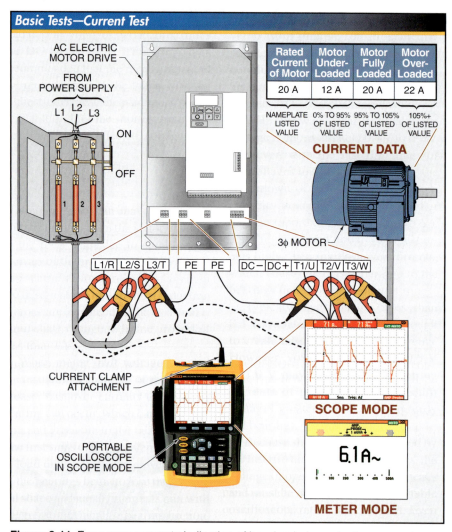

Basic Tests—Current Test

AC ELECTRIC MOTOR DRIVE

FROM POWER SUPPLY

L1 L2 L3

ON

OFF

1 2 3

Rated Current of Motor	Motor Under-Loaded	Motor Fully Loaded	Motor Over-Loaded
20 A	12 A	20 A	22 A
NAMEPLATE LISTED VALUE	0% TO 95% OF LISTED VALUE	95% TO 105% OF LISTED VALUE	105%+ OF LISTED VALUE

CURRENT DATA

3φ MOTOR

| L1/R | L2/S | L3/T | PE | PE | DC − | DC+ | T1/U | T2/V | T3/W |

CURRENT CLAMP ATTACHMENT

PORTABLE OSCILLOSCOPE IN SCOPE MODE

SCOPE MODE

6.1A~

METER MODE

Figure 3-11. For a more accurate indication of how loaded a motor is, a current measurement must be taken.

A portable oscilloscope is one of the best tools to use for taking advanced measurements because of the advanced test functions. For example, once an advanced test indicates a potential problem, such as high transient voltages at a motor or waveform distortion, a portable oscilloscope can be used to isolate the problem. A portable oscilloscope can determine if the high transient voltages are caused by connection wires that are too long between the motor drive and the motor leads and if the motor drive's carrier frequency is not set correctly for the application. **See Figure 3-12.**

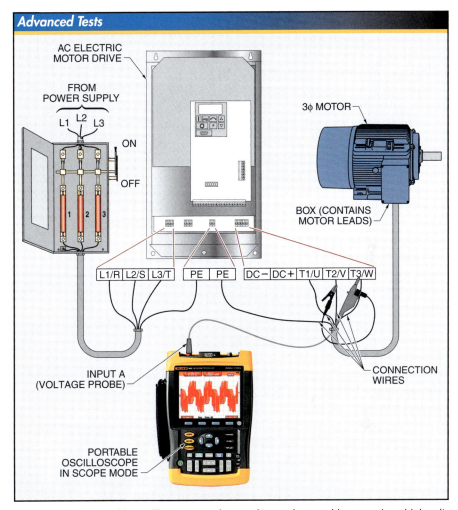

Advanced Tests

AC ELECTRIC
MOTOR DRIVE

FROM
POWER SUPPLY

L1 L2 L3

ON

OFF

3φ MOTOR

BOX (CONTAINS
MOTOR LEADS)

| L1/R | L2/S | L3/T | | PE | PE | | DC− | DC+ | T1/U | T2/V | T3/W |

INPUT A
(VOLTAGE PROBE)

CONNECTION
WIRES

PORTABLE
OSCILLOSCOPE
IN SCOPE MODE

Figure 3-12. A portable oscilloscope can be used to analyze problems such as high voltage spikes between the motor drive and the motor leads.

Diagnostic Tests

Basic and advanced test measurements can determine the most common problems and also provide an indication of more serious and difficult to locate problems that can damage equipment over time, such as poor power quality. For system and equipment diagnostics and documentation, such as locating power quality problems, advanced measurements must be taken and closely reviewed. While advanced DMMs and power quality meters can help perform some diagnostic testing and can identify most problems, a portable oscilloscope is the best instrument for diagnostic testing and documentation because of its advanced functions, high sampling rate, and waveform display.

CASE STUDY: HVAC Service Call

Equipment to Be Serviced: A call was received about a standard commercial rooftop HVAC unit that controls one of several zones within the building.

Service Issue: The HVAC unit has malfunctioned as observed by building occupants reporting uncomfortable temperatures.

Past Service and Repair Reports: Before servicing the HVAC unit, maintenance personnel reviewed the unit's previous service and repair reports. The reports indicated that the HVAC unit operated without problems for several years after installation, but had a failure three months prior due to a reported fan-motor failure. The fan motor was replaced and the HVAC unit was returned to operation.

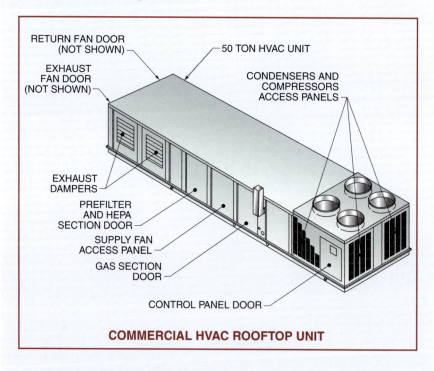

RETURN FAN DOOR (NOT SHOWN)

EXHAUST FAN DOOR (NOT SHOWN)

50 TON HVAC UNIT

CONDENSERS AND COMPRESSORS ACCESS PANELS

EXHAUST DAMPERS

PREFILTER AND HEPA SECTION DOOR

SUPPLY FAN ACCESS PANEL

GAS SECTION DOOR

CONTROL PANEL DOOR

COMMERCIAL HVAC ROOFTOP UNIT

Applied Troubleshooting Procedure

1. The circuit breakers in the service panel feeding the unit were inspected before going to the roof to test the HVAC unit. The circuit breakers were not tripped.

2. The main disconnect switch on the outside of the HVAC unit was opened so the power supply and fuses could be inspected. A voltage measurement taken with a standard DMM indicated that the main supply voltage was within the HVAC unit's operating range and the fuses were good. The power was turned OFF and locked out so that the HVAC unit could be opened and visually inspected without the power ON. A visual inspection indicated no problems.

3. Power was restored to the unit so that additional measurements could be taken.

   • The variable frequency drive (VFD) display was ON and indicated no fault codes, which could have indicated any one of several faults (under- or overvoltage, motor overloaded, ground fault, VFD over temperature, etc.).

   • The VFD display indicated that the fan motor should be ON, although it was not. This indicates the main problem is that without the fan motor on, the unit would soon automatically shut off to prevent additional problems from overheating.

   • Power to the HVAC unit was turned OFF.

4. The fan motor was visually inspected and there were no signs of overheating, wear, or damage. The motor connectors where exposed allowing voltage measurements to be taken at the motor.

   • Power was restored to the HVAC unit, and voltage at the fan motor was measured.

VOLTAGE
READING AT
FAN MOTOR ———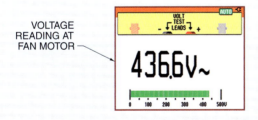

   • The voltage at the fan motor was within its nameplate rating, but the fan motor was not operating, which was unusual since the fan motor had been replaced within the previous three months.

   • Power to the HVAC unit was turned off so that the fan motor could be replaced.

5. The nameplate information was recorded from the failed fan motor so a replacement motor could be purchased. When reviewing the past service and repair report, it was indicated that a different motor than the OEM's motor was ordered to save downtime. A replacement motor with the exact frame size, horsepower rating, and voltage rating was ordered.

6. The new fan motor was installed, and the power was turned back ON. The HVAC unit operated with no problems.

7. The reported problem has been corrected and a service and repair report could be completed, ending the assigned maintenance task. The fact that the replaced motor only operated for three months before it failed could be explained by the fact that components can prematurely fail due to a manufacturer defect, damage during installation, or improper design flaws. However, it is important to troubleshoot the system for proper operation and to correct any secondary problems when a component is returned to operation.

Tests that can be performed include the following:

- Measure the voltage at the main power supply and motor to determine any voltage drop that could be caused by loose connections or undersized conductors.

- Measure the current draw of the motor and total equipment and compare the readings with the listed nameplate ratings. Current at the listed nameplate rating indicates that the component is operating at 100%. Current above the listed nameplate rating indicates that the component is overloaded. Higher current draw can be a sign of dirty filters or coils, undercharged refrigerant, belts that are too tight, or misalignment.

- Measure voltage and current over time to determine how the component is operating during each cycle or condition. Use the MIN/MAX recording function and allow the meter to record for at least one complete operation cycle.

MOTOR
CURRENT
DRAW —

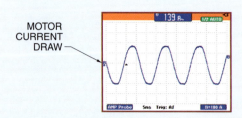

If additional tests and measurements indicate signs of improper operation, additional investigative and diagnostic tests and measurements should be performed. Investigative tests include comparing the replacement part's specifications to the original part's specifications and verifying with the OEM that there is not a weak point in the system. Noticeable problems, such as an undersized HVAC unit, poor placement of an HVAC unit causing restricted airflow (such as plants that have grown around the unit), and human error (such as unqualified individuals working on or near the unit), should be searched for.

Diagnostic testing is the most advanced form of troubleshooting, but does not have to be difficult or only apply to intermittent or complex problems. Diagnostic testing and the documentation produced must be a regular activity in preventive maintenance and service calls. Diagnostic testing is the best method to prevent future problems and provides the information required to take corrective action.

8. While performing investigative testing, it was discovered that although the replacement motor has the same horsepower and voltage rating as the original motor, the insulation rating on the replacement motor is Class A (low grade) while the original motor insulation was Class H (high grade). Different insulation classes could be the cause of the problem since the VFD drives the motor, and high transient voltages at the motor destroy the motor's insulation.

9. Before the premature motor failure is attributed to low-grade insulation, diagnostic testing should be performed. A portable oscilloscope is used to measure, display, and record different operating factors such as voltage (including peak), current, frequency, true power, and apparent power. The portable oscilloscope displays the high voltage peak at the motor, which is often caused by conductor cable lengths that are too long between the VFD and motor or a high carrier frequency setting. The observed and recorded high transient voltages determine that the motor problems are caused by the Class A insulation. *Note:* Refer to Chapter 4 for more information about high voltages when troubleshooting variable frequency drives.

PEAK VOLTAGE
MEASUREMENTS

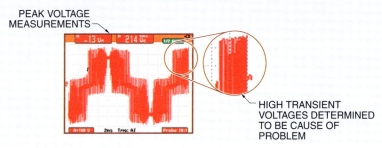

HIGH TRANSIENT
VOLTAGES DETERMINED
TO BE CAUSE OF
PROBLEM

PWM
VOLTAGE

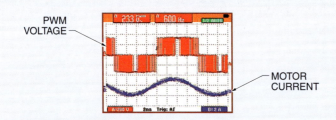

MOTOR
CURRENT

Repair Solutions

Based on the diagnostic test and measurement results, the following solutions are recorded in the service and repair report:

1. To immediately help protect the motor from high transient voltages, the VFD frequency is reduced to 5 kHz from 16 kHz, resulting in lower transient voltages at the motor. The portable oscilloscope measurements at the motor verified lower transient voltages.

2. There is a high probability that a replacement motor with Class A insulation can prematurely fail again, since it is not designed for the application. A replacement from the OEM (or a replacement motor with Class H insulation) must be ordered.

Maintenance Recommendations

Based on the information provided during the service call, the following recommendations are documented:

1. Before securing a replacement motor or component that is not from the OEM, verify that the replacement motor or component meets each specification.

2. While performing service calls, measured electrical quantities and portable oscilloscope screen captures must be recorded as hard copy and placed in a file included with the paperwork located with the equipment for future reference and comparisons.

4

Motor and Drive Troubleshooting

Electrical systems and equipment must be designed, selected, and installed so that they can safely and efficiently perform the required work. Once installed, electrical systems and equipment must be maintained so that they operate as designed with the least amount of problems possible. When electrical systems and equipment do not operate as designed, troubleshooting procedures must be performed to locate and correct the problem.

TROUBLESHOOTING

Troubleshooting is the systematic elimination of the various parts of a system or process from consideration during the process of locating a malfunctioning part. *Preventive maintenance* is scheduled work required to keep equipment in peak operating condition. *Energy efficiency (green technology)* is the process of selecting, replacing, or modifying equipment and systems to reduce the amount of power used and still provide the same amount of service.

Energy efficiency requires selecting products that are safe for the environment and as sustainable as possible. Selecting equipment that is safe and sustainable and regular preventive maintenance help reduce electrical motor and drive problems, which minimizes the need for troubleshooting.

Locating and correcting an electrical motor or drive problem requires experience with the equipment and system, knowledge of the required test instruments, and access to logical procedures. With each troubleshooting task performed, experience concerning the equipment, system, and applicable test instruments is gained and can be used when performing future troubleshooting tasks. Documenting troubleshooting tasks produces a record that can be referenced when similar problems occur in the future. Using test instruments such as portable oscilloscopes allows for easier documentation of problems and their solutions.

Troubleshooting Using Experience

Troubleshooting using experience is a method of locating a malfunction in a system or process by applying information learned from reading the original equipment manufacturer's (OEM's) manuals, attending training classes and seminars, experiences with similar malfunctions, and experience gained each time electrical components, equipment, and systems are tested. Knowledge gained from previous information and experience can be applied to future troubleshooting tasks.

Troubleshooting Using Manufacturer's Documentation

The OEMs of motors and motor drives usually include various preventive maintenance and troubleshooting suggestions, along with installation directions and other relevant documentation, when their products are acquired. Relevant documentation includes documents such as installation manuals, user's manuals, and operation manuals.

Documentation is usually provided with the equipment or can be accessed through the Internet. Information provided by an OEM typically includes guidelines designed to help repair the most common problems associated with the equipment. For example, a typical motor drive troubleshooting manual from the OEM provides corrective action to apply when a specific problem occurs. **See Figure 4-1.**

OEM Motor Drive Troubleshooting Manuals

TABLE 9 - TROUBLESHOOTING

INDICATION	POSSIBLE CAUSE	CORRECTIVE ACTION
No LED indication	Lack of input voltage	Check input power
	Loose connections	Check input power termination
		Measure voltage into drive for correct level
Motor will not start	Not enough starting torque	Increase torque boost parameter setting
	Motor overloaded	Correct motor loading
		Resize motor/inverter system
Motor will not reach maximum speed	Maximum frequency limit adjusted too low	Adjust maximum frequency limit parameter setting
	Motor overloaded	Check for mechanical overload. If unloaded motor shaft does not rotate freely – replace motor bearings
	Speed potentiometer failure	Replace potentiometer

Figure 4-1. Motor drive troubleshooting manuals from the original equipment manufacturer (OEM) provide corrective action to apply when a specific problem occurs.

Use a properly rated clamp meter when measuring motor loads.

As with any electronic device, motor drives can develop problems. The problem may be with the motor drive, components connected to the drive, or with the power distribution system. After installing the motor drive or prior to performing preventive maintenance and troubleshooting procedures, the equipment, system, and immediate work area should be inspected for any indications of potential problems. Actions that can be taken to check for problems include the following:

- Verify that the area near the motor drive enclosure is clean and that all warning labels and precautions are properly located and followed.
- Verify that components, devices, and conductors are connected and that the motor and motor drive is properly vented.

- Verify that motor is properly rated for the drive application.
- Inspect equipment for excessive moisture, dirt, corrosion, burning, charring, and any other unusual conditions that could affect equipment and system operation.

Troubleshooting Using Internal Procedures

Most commercial and industrial facilities have written internal procedures for troubleshooting equipment. Internal procedures typically ensure the personal safety of individuals and maintain records, rather than applying specific troubleshooting tasks to locate a problem. It is important that facility procedures are properly understood and followed. A supervisor or other qualified individual should be consulted if a procedure is not understood by personnel working on or near the equipment.

TECH TIP

Test instruments are a direct link between a live circuit and your body. Use only test instruments that are in good condition and rated for the measurement and area to be used.

Troubleshooting Checklists

Troubleshooting checklists are used by technicians, electricians, and other personnel to gather information that can be helpful when trying to locate a problem and can be used to determine secondary problems. Checklists also provide a form of written documentation concerning specific equipment, systems, types of problems, and corrective actions taken. **See Figure 4-2.**

Troubleshooting Checklists . . .

MOTOR and DRIVE TROUBLESHOOTING CHECKLIST

Problem Observed or Reported:
- ☐ System not ON/running
- ☐ System not operating normal or as required
- ☐ System making noise
- ☐ System causing electrical shocks
- ☐ Other _____

Power Distribution Type:
- ☐ 1φ
- ☐ 3φ Y
- ☐ 3φ Δ
- ☐ Fuses
- ☐ Circuit breakers
- ☐ Circuit voltage(s) rating: ___V
- ☐ Circuit amperage rating: ___A

Motor Type:
- ☐ Motor manufacturer/type ___
- ☐ Motor power/voltage/current rating ___ HP/kW, ___V, ___A, ___SFA

Drive Type:
- ☐ Drive manufacturer/type ___
- ☐ Drive maximum power rating ___
- ☐ Displayed drive faults code ___

Problem Pattern:
- ☐ Day(s) of week:
- ☐ First time
- ☐ Recent repeat
- ☐ Random
- ☐ Monday
- ☐ Tuesday
- ☐ Wednesday
- ☐ Thursday
- ☐ Friday
- ☐ Saturday
- ☐ Sunday
- ☐ Time(s)
- ☐ Reported time = ___ or ☐ Morning ☐ Afternoon ☐ Evening ☐ Night
- ☐ Random repeat
- ☐ Always same approximate time

Problem History:
- ☐ Has problem been previously observed or reported? ☐ No ___ ☐ Yes ___
- ☐ Was corrective action taken? ☐ No ___ ☐ Yes ___
- ☐ Are other parts of the system affected? ☐No ___☐ Yes ___
- ☐ If "yes", list here: ___
- ☐ Has there been any recent work changes made to the system? ☐ No ___ ☐Yes ___

Possible Problem(s):
- ☐ Blown fuses/open circuit breakers
- ☐ Drive automatically tripped OFF
- ☐ Control circuit problem
- ☐ Operator problem

. . . Troubleshooting Checklists

Required PPE for expected test/measurements:
- ☐ Safety glasses
- ☐ Arc shield
- ☐ Electrical gloves (leather & rubber)
- ☐ Head protection
- ☐ Ear Protection
- ☐ Protective rated clothing (fire resistant)
- ☐ Rubber insulated matting
- ☐ Other_____

Minimum required CAT rating of test instruments, test leads, and attachments:
- ☐ CAT I
- ☐ CAT II
- ☐ CAT III
- ☐ CAT IV

Expected test instruments required:
- ☐ DMM - Type ___ Attachments ___
- ☐ Clamp-on ammeter
- ☐ Power quality analyzer
- ☐ Portable oscilloscope
- ☐ Insulation tester
- ☐ Noncontact IR thermometer
- ☐ Thermal imaging camera

Numerical Measurements Taken:
- ☐ Voltage into drive ___V
- ☐ Current into drive ___A
- ☐ Voltage to motor from drive ___V
- ☐ Motor current ___A
- ☐ Motor power ___ kW, ___VA, ___PF
- ☐ Other ___

Scope Waveforms/Measurements Taken/Observed:
- ☐ Voltage into drive. Observations: _____
- ☐ Current into drive. Observations: _____
- ☐ Motor voltage. Observations: _____
- ☐ Motor current. Observations: _____
- ☐ Motor power. Observations: _____
- ☐ Transients at motor = _____
- ☐ Other _____

Problems or Potential Problems Found:
- ☐ Fuses
- ☐ Circuit breakers
- ☐ Drive parameters not set correctly
- ☐ Drive not working correctly
- ☐ Drive working correctly, equipment/motor not working
- ☐ Motor problem
- ☐ Other problem (_____)

Figure 4-2. Troubleshooting checklists are used to gather information that can be helpful when trying to locate a problem and can help determine secondary problems.

ASI Robicon

Motors and motor drives produce work in constant horsepower loads such as those found in paper mills.

COMMON MOTOR AND DRIVE PROBLEMS

Electric motors and motor drives are used to produce work in residential, commercial, industrial, entertainment, government, and other applications that require a controlled rotating output. Problems in such applications typically include common problems that occur in most electrical systems and certain less common problems associated with the application requirements, control requirements, and environmental conditions. Common problems that can occur in most motor and drive applications include the following:

- blown fuses and tripped circuit breakers
- damaged equipment from moisture, dirt, corrosion, lightning, mishandling and any other condition the equipment is not designed to withstand

- damage from age, over usage, under sizing, misalignment, and improper power conditions such as loss of phase, transient voltages, harmonics, and voltage sags and swells

Motor Problems

When electric motors are properly installed, sized, and connected to a power source, they are one of the most reliable pieces of electrical equipment. It is not uncommon for electric motors to operate for ten years or longer with little or no maintenance. However, because motors can and do fail, OEMs and end users have tracked causes of motor failures for over 100 years, providing documentation of causes for common motor malfunctions and failures. **See Figure 4-3.**

Single-Phasing

Single-phasing results from the total loss of one of the phase voltages applied to a three-phase (3ϕ) AC induction motor and can be difficult to detect under normal operation. In a variable speed drive (VSD) application, single-phasing is usually caused by an open connection at either end of the cabling between the motor and drive or in one of the cable conductors. It is also possible that an insulated gate bipolar transistor (IGBT) could have failed as an open circuit, although some VSDs can detect and signal this condition internally. An *insulated gate bipolar transistor (IGBT)* is a solid-state, three-terminal device that is used for fast switching of electric circuits.

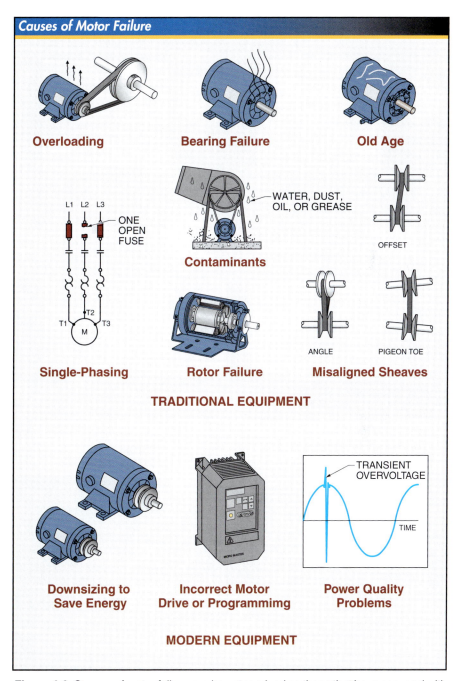

Causes of Motor Failure

Overloading

Bearing Failure

Old Age

Single-Phasing

Contaminants

WATER, DUST, OIL, OR GREASE

OFFSET

Rotor Failure

Misaligned Sheaves

ANGLE

PIGEON TOE

TRADITIONAL EQUIPMENT

Downsizing to Save Energy

Incorrect Motor Drive or Programmimg

TRANSIENT OVERVOLTAGE

TIME

Power Quality Problems

MODERN EQUIPMENT

Figure 4-3. Causes of motor failure can be categorized as those that have occurred with traditional equipment or those that have occurred only with more modern equipment.

Single-phasing is a fairly common cause of failure for 3φ induction motors. When single-phasing occurs, the other two phase windings must conduct more current, which produces more heat and leads to premature motor failure. Single-phasing may not be detected because the motor continues to run, although with increased heat and possibly a loss of torque. Additionally, voltage measurements made at the motor terminals read close to normal because motor action induces a voltage into the open winding.

For this reason, taking current measurements on all phases until an open phase is detected is the best method for detecting single-phasing. Another indication of a possible single-phasing condition is that if the motor is stopped and restarted, it may run backwards.

TECH TIP

The causes of single-phasing include blown fuses, mechanical failures within the switching equipment, and lightning strikes that knock out power lines.

Overvoltage Reflections

Fast rise times on the VSD output pulses and long cable runs between the VSD and the motor can cause overvoltage reflections more than twice as high as the DC bus voltage. A reflection-induced overvoltage condition is hazardous because it can damage motor windings over time. A portable oscilloscope is required to troubleshoot this problem. **See Figure 4-4.** Because there is only a small V_{rms} difference between the two waveforms, this problem cannot be detected with a voltmeter or DMM.

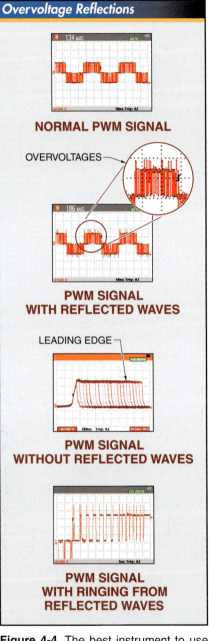

Overvoltage Reflections

NORMAL PWM SIGNAL

OVERVOLTAGES

PWM SIGNAL
WITH REFLECTED WAVES

LEADING EDGE

PWM SIGNAL
WITHOUT REFLECTED WAVES

PWM SIGNAL
WITH RINGING FROM
REFLECTED WAVES

Figure 4-4. The best instrument to use to detect the overvoltage conditions that result from reflected voltages is a portable oscilloscope.

A reflection-induced overvoltage condition may not show up as a problem when the PWM drive is first installed. After the problem has been identified, the simplest solution for an overvoltage or ringing problem is to shorten the cable. **See Figure 4-5.** However, if the cabling in the PWM application cannot be shortened, the problem can be corrected through one of the following actions:

• installing an external low pass filter between the VSD output terminals and the cable to the motor to slow the rise time of the PWM signal

• applying series line reactors between the VSD output terminals and the cable to the motor (small horsepower applications only)

• installing an RC-impedance matching filter at the motor terminals to minimize the overvoltages or ringing effects

Note: All the solutions should be designed for specific applications by a qualified engineer.

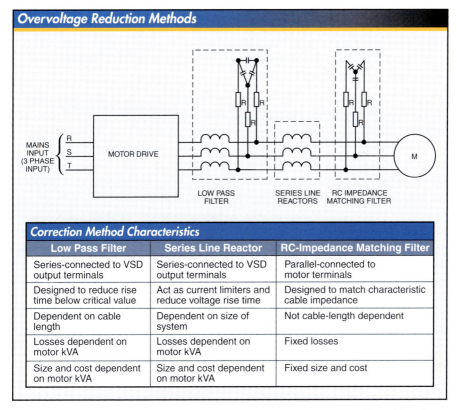

Overvoltage Reduction Methods

Correction Method Characteristics		
Low Pass Filter	**Series Line Reactor**	**RC-Impedance Matching Filter**
Series-connected to VSD output terminals	Series-connected to VSD output terminals	Parallel-connected to motor terminals
Designed to reduce rise time below critical value	Act as current limiters and reduce voltage rise time	Designed to match characteristic cable impedance
Dependent on cable length	Dependent on size of system	Not cable-length dependent
Losses dependent on motor kVA	Losses dependent on motor kVA	Fixed losses
Size and cost dependent on motor kVA	Size and cost dependent on motor kVA	Fixed size and cost

Figure 4-5. Reflection-induced overvoltage problems can be corrected by adding filters or series line reactors to the circuit.

WARNING: Reflected voltage may lead to peak voltages two to three times the DC bus voltage. Therefore, it is recommended that the measurement at the motor terminals be taken with the highest rated test measurement probe available and for the shortest time possible, wherever reflected voltages are likely to be present.

Bearing Currents

In a motor, there is an unavoidable shaft voltage created from the stator winding to the rotor shaft due to small asymmetries of the magnetic field in the air gap. These asymmetries are inherent in the design of the motor. Most induction motors are designed to have a maximum shaft voltage to frame ground of less than 1 V_{rms}.

When motor shaft voltages exceed the insulating capability of the bearing grease, flashover currents to the outer bearing can occur, which can cause pitting and grooving to the bearing races. The first signs of this problem are usually loud unusual noise and overheating, as the bearings begin to lose their original shape and metal fragments mix with the grease and increase bearing friction. This can destroy bearings within a few months of operation of the VSD, causing costly electric motor repairs and downtime.

TECH TIP

Most electric motor manufacturers have developed shaft grounding systems to eliminate the problem of bearing currents.

Motor shaft voltages may also develop from internal, electrostatically coupled sources, including belt-driven couplings and rotor fan blades when air passes over them such as in steam turbines. When motor supply voltage is a 60 Hz sine wave, bearing breakdown voltage is approximately 0.4 V to 0.7 V. However, with the fast rise times of the voltages found of PWM drives, the breakdown of the grease's insulating capacity actually occurs at a higher voltage (approximately 8 V to 15 V). This higher breakdown voltage creates higher bearing flashover currents, which causes increased damage to the bearings in a shorter amount of time.

Typically, shaft voltages of less than 0.3 V are safe and are not high enough to create destructive bearing currents. However, voltages from about 0.5 V to 1.0 V may create harmful bearing currents of greater than 3 A (shaft voltages greater than 2 V can destroy the bearing).

Because shaft voltages are caused by the fast rise times of the drive pulses, the voltages appear as inconsistent peaks and must be measured using a portable oscilloscope rather than a DMM. Even if the DMM has peak detect, there is enough variation between peaks to render the reading unreliable.

A dedicated shaft voltage probe can help take shaft voltage measurements more safely. A dedicated shaft voltage probe is similar to a small conductive brush that connects to a portable oscilloscope probe tip. It can be used to establish the test connection with a motor shaft surface.

A shaft voltage probe is an extension rod that extends the probe body so that measurements can be taken at a safer distance from the point of work. The common lead is connected to the motor's frame for grounding purposes. Shaft-to-frame ground voltage measurement must be taken after the motor has warmed to its normal operating temperature because shaft voltages may not be present when the motor is cold.

TECH TIP

The best solution for a bearing current problem is to reduce the carrier (pulse) frequency of the drive to a range of 4 kHz to 10 kHz. If the carrier frequency is already in this range, alternative solutions can be applied such as the use of shaft grounding devices, bearing insulation, a Faraday shield in the motor, conductive grease, ceramic bearings, or filtering between the VSD and the motor.

Leakage Currents

The amount of leakage current that results from capacitive coupling between a motor's stator winding and frame ground increases with PWM drives as the result of the higher switching frequencies used, which worsens the problem. The increase in leakage currents can also cause nuisance tripping of ground fault protection relays. A common mode choke along with a damping resistor can be used to reduce leakage currents. Also, special EMI suppression cables can be used between the drive output and the motor terminals. Isolation transformers on the AC inputs can also reduce common mode noise.

Induced Electrical Noise

Induced electrical noise can significantly affect sensitive control circuits (e.g., speed, torque, control logic, and position feedback sensors) as well as outputs to display indicators and system control computers. Because many control inputs are scaled 0 V to 5 V (or 10 VDC maximum) with typical resolutions of one part in a thousand, just a few millivolts of induced noise can cause improper operation. Significant amounts of noise can actually damage a drive and/or motor.

Other common sources of electrical noise are relay and contactor coils. Transient voltages caused by the opening of the coil circuits can generate spikes of several hundred volts, which can induce several volts of noise in adjacent wiring. Good installation practices should be followed by using twisted-pair, shielded wiring for sensitive control circuits. The twisted-pair, shielded wiring should be separated from the relay and contactor coil circuit wiring. Adding snubber circuits to the relay and contactor coils can reduce arcing and the noise induced in adjacent wiring.

Noise on the line inputs from SCR-controlled DC drives, current source inverters, six-step drives, and other high-noise loads in a building can also induce unwanted noise in adjacent control wiring. The high-energy, fast-switching PWM signals on motor cabling can also contribute to this problem if the cabling is unshielded and close to control wiring. The best method for minimizing this problem is to verify that line input wires and motor cabling are contained in separate grounded, rigid metal conduit.

Motor drives are typically installed in a separate box near the motors they control.

Determining if noise problems exist in control circuit wiring requires a portable oscilloscope because it can capture and display noise and transient events that a DMM cannot. An oscilloscope with multiple channels further enhances troubleshooting abilities by enabling the simultaneous visualization of multiple signals. Signals may also be viewed simultaneously side by side as both current and voltage.

Special care must be taken when using a portable oscilloscope to take low voltage measurements so that noise is not coupled into the oscilloscope and then mistaken as noise on the control signal wiring. Using probes with short ground leads can minimize noise introduced by the oscilloscope probes into the measurement.

Volts/Hertz (V/Hz) Ratio

The ratio of voltage to frequency determines the amount of torque produced by an AC induction motor. If the motor has a loss of torque, the V/Hz ratio should be measured. Portable oscilloscopes can simultaneously display the frequency of the PWM output and a voltage comparable to the motor nameplate rating. By measuring the V/Hz ratio, problems with the following components can be revealed:

- DC bus circuits—stable frequency with low, high, or unstable voltage
- IGBT control circuits—unstable frequency with voltage in specified range
- speed inputs to control boards—voltage and frequency are fluctuating together, or the speed of the motor is off but the V/Hz ratio is correct

Inverter Drive Circuits

Even though PWM drives have replaced voltage source inverters (six-step drives), there are still voltage source inverters in operation that require maintenance. Common problems that are present in six-step drives are voltage and current imbalance, single-phasing, and overheating. Less comon problems for this type of drive include shorted transistors.

A shorted transistor on certain six-step drives can be detected by measuring across the transistor with a portable oscilloscope. On a portable oscilloscope screen, a properly operating transistor has a square waveform with sharp edges, while a malfunctioning transistor produces waveforms that are rounded at the peak of the leading edge.

If the shorted transistor is causing the drive's protection circuit to trip, then the converter section that rectifies the AC into DC can be disconnected. The inverter circuit can then be run with the 10 V of leakage voltage that is present on the DC bus. The input driver circuit can switch on the transistors but at a lower voltage level, and the malfunctioning transistor can be easily detected. Also, the inverter section can be disabled while troubleshooting the AC-to-DC converter circuit. The speed control can be varied while monitoring the DC bus voltage to verify that it varies with the speed control.

Note: Voltage feedback resistors must remain connected to the DC bus to ensure that the converter section is controlled by the speed potentiometer. Disconnect the AC inverter section after the voltage feedback resistors. If this procedure is not followed, the DC converter can immediately switch on when the drive is started.

PWM Inverters

Most modern fractional-horsepower PWM drives are integrated to the point where the input diode block and IGBTs are potted into a single disposable module. The module is bolted to the heat sink.

If a phase output to the motor is missing or partially conducting, the input section of the IGBT can be tested for signal integrity. If the inputs

are within proper specifications (compared with a known good unit), then the IGBT block can be replaced. If the inputs are not within manufacturer's specifications, the IGBT block could be loading the input drive circuit or the IGBT drive board.

If inputs to the IGBTs are not operating properly, it must be determined whether the problem is with the circuit board supplying inputs to the IGBT block or the IGBT block itself. If accessible, the IGBT should be tested in-circuit with the power off and isolated from the motor and IGBT drive PCB by removing the connections. The diode test function of a DMM should be used to test the IGBT. **See Figure 4-6.**

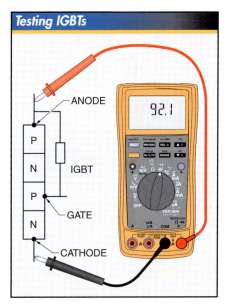

Figure 4-6. The diode test function of a DMM is used to test an insulated gate bipolar transistor (IGBT).

Transients and Swells

The most common causes of tripping of the overvoltage circuit fault on VSD inverters are motor regeneration and transients and swells on the AC line inputs. To test for these conditions, a portable oscilloscope or a power line monitor with at least 10 μsec/division resolution that can time-stamp an event when it occurs should be used.

The portable oscilloscope can measure and time-stamp transients so that they can be time-correlated to whatever event caused the VSD fault. Alternatively, power quality analyzers can be used to find disturbances in the three-phase power supply. These instruments can monitor four voltage lines (three phases plus neutral) and four currents simultaneously to provide details about abnormal readings. The details provided include the voltage or current changes, frequency changes, transients, etc. with time stamps for each event.

If circuit fault tripping is caused by transients, an isolation transformer or series line reactor can be placed in series with the front end of the VSD. An alternate solution is to place a surge protection device (SPD) at the motor control center or at the primary side of the distribution transformer feeding the VSD. However, if the source of the transient is another load on the same secondary feed as the VSD, then a separate isolation transformer or series line reactor may need to be used directly in front of the VSD or the VSD may need to be connected to its own feed.

If a motor drive is installed in an area where lightning occurs often, it should be verified that the building has proper surge protection. Additionally, the building's grounding system must be properly installed and functioning to help dissipate lightning strikes safely to ground rather than through some path in the building's power distribution system.

Voltage swells can be monitored using a line monitor such as a portable oscilloscope. To offset the effects of a voltage swell, a temporary dropout relay can be installed that is active for as many cycles as the voltage swell and that can be tolerated by the motor drive. The viability of this solution is determined by the amount of "ride-through" the VSD's input circuit has, that is, the length of time before the DC bus voltage drops to an undervoltage condition.

Typically, a building that is deficient in proper wiring and grounding is susceptible to transients, sags, and swells. Transients, sags, and swells on electrical and electronic equipment should therefore be treated as possible symptoms of underlying systemic problems that should be corrected.

TECH TIP

Voltage swells can be more destructive than voltage sags because it generally takes less time to damage hardware with a higher voltage than with a lower voltage. Even a very short high-voltage condition can cause permanent equipment or component damage.

Motor Regeneration

A common source of overvoltage on a DC bus is motor regeneration. *Motor regeneration* is a condition that occurs in an electric motor when a load is coasting and begins to drive the motor rather than being driven by it. This causes the motor to function as a generator that returns energy into the DC bus rather than drawing energy from it. Motor regeneration can be measured by testing for a change in the direction of the DC current into the DC bus while simultaneously checking the DC bus voltage for an increase above the trip point.

Dynamic braking is a method of motor braking in which a motor is reconnected to act as a generator immediately after it is turned off. If regeneration is causing the overvoltage tripping, dynamic braking can be used to limit how fast the regenerative current is allowed to feed back into the DC bus capacitors. If dynamic braking has already been used but does not function properly, it can be tested to verify that it is working according to the manufacturer's specifications. Other considerations of dynamic braking include the following:

• A resistor dynamic brake should be inspected for signs of overheating, such as discoloration, cracking, or the smell of overheated components. The resistance value can also be measured against the manufacturer's specifications.

• Transistor dynamic brake silicon junctions can be tested using a DMM diode test function. Also, braking current can be measured and the current waveform compared with that of a properly functioning system.

Low Voltage

There are several causes of nuisance tripping of the low-voltage fault circuit

on VSD inverters. Voltage sags and undervoltage conditions on the line input to the drive are common conditions associated with nuisance tripping. Sags and undervoltage conditions are often caused by another load being turned on within the building's distribution system or possibly by the starting of a large electrical load by an adjacent building. A portable oscilloscope can be used to diagnose low voltage conditions by taking measurements that can time-stamp the sag or the event when the undervoltage causes the VSD to trip with a low-voltage fault. **See Figure 4-7.**

Measurements should be taken at the service entrance in order to determine if the sag is originating within or outside the building. Voltage and current should be monitored simultaneously to determine if the problem is downstream from the service entrance. An upstream problem has a voltage sag without a corresponding surge in current or with the voltage and current of the same polarity. If the problem is within the building, measurements should be continued at different load centers until the load has been isolated with the corresponding voltage sag and current surge.

A motor that is drawing enough current to cause the DC bus voltage to drop below the undervoltage fault setting, but not enough to trip the current overload, can also cause this condition. The motor current should be tested for overloading. Also, it should be verified whether the program settings of the drive are correct for the motor nameplate ratings, including the application for which the motor and drive were intended.

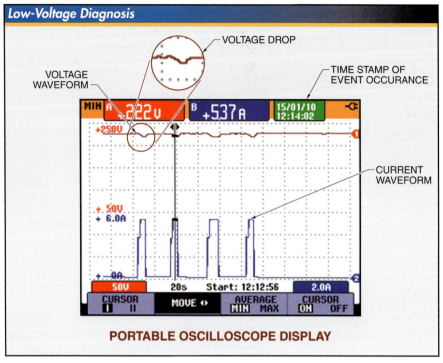

Figure 4-7. With voltage sags caused by a downstream load, voltage decreases every time the load current increases.

The line input voltage waveform to the VSD must be a well-formed sine wave. Severe flat-topping of the waveform can prevent the DC bus capacitors from charging fully, which lowers the DC bus voltage and the amount of current available to the VSD output circuit. **See Figure 4-8.**

Diode Bridges

A *diode bridge* is an electrical circuit that is used to convert AC into DC. It contains four diodes that permit both halves of input alternating current sine waves to pass. Diode bridges used in PWM drives are easily tested. The normal failure mode for a diode is from transient overvoltages

or overcurrent conditions. If a shorted diode trips a circuit breaker before it has a chance to burn into an open circuit, then a DMM with a diode test function can locate this problem.

A diode test function is used because the resistance measurement function may not produce a voltage high enough to get the diode to conduct. When power has been completely disconnected from the VSD line inputs, the diode test function should be used to test from both the negative and positive DC bus to each of the line input connections. Each line input should be probed with the negative lead, starting with the positive lead on the +DC bus.

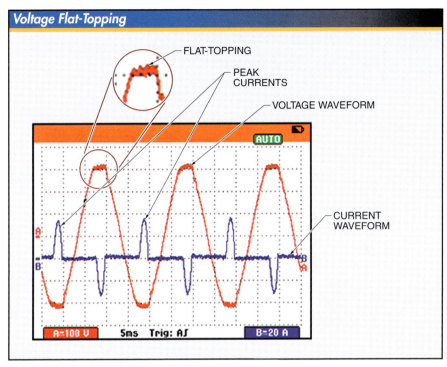

Figure 4-8. Voltage flat-topping can be caused by peak currents with high levels of harmonics.

Each reading should indicate "OL" (overload) or a reverse bias condition. The measurements should be taken again, only now with the negative lead on the +DC bus and with each line input measured with the positive lead. This will forward-bias the diodes and cause them to conduct with about 0.5 V to 0.7 V dropped across them. This same procedure should be used for the other three diodes connected to the negative DC bus. Shorted diodes have a reading of 0 V in both directions. Open diodes read overloads in either direction when they should be conducting.

Most modern VSDs also have a precharge circuit to reduce inrush

current to the DC bus capacitors and prevent tripping of the protection circuits. Motor drives less than 200 HP (150 kW) have a precharge resistor that limits the inrush current until the DC bus charges to about 60%, after which a relay removes the resistor from the circuit. When troubleshooting the input circuit, it should be verified that the precharge resistor and relay are not overlooked as possible causes of the problem. VSDs with horsepower greater than 200 HP usually have SCRs that slowly charge the DC bus capacitors until the DC bus voltage rises to a predetermined level, when the SCRs go into full rectification.

Voltage Notching

A *voltage notch* is a fast switching disturbance of the normal voltage waveform. Six-step drives usually use SCRs rather than diodes, such as those used in PWM drives, to rectify the input line voltage and convert it to DC.

Line voltage can be tested with a portable oscilloscope for distortion, as seen by notches in the display of the sine wave. The distortion is caused by the firing of SCRs from the control circuit. **See Figure 4-9.** When distorted voltage enters the distribution system and is applied to other sensitive electronic loads, equipment can turn on and off at the wrong times, which can cause damage to the equipment and related systems.

The most common way to correct voltage notching from six-step drives is to connect line reactors in series with the line input to the drive, or use an isolation transformer. The advantage of an isolation transformer is that it also reduces common mode noise.

TECH TIP

Isolation transformers act as a buffer among a nonlinear power supply, the load, and the power source. They are used to provide clean power to nonlinear loads and to prevent harmonics from moving upstream into the power source.

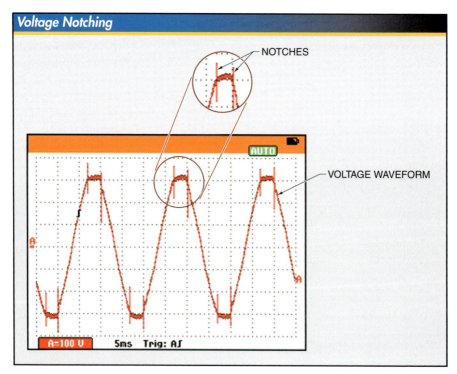

Figure 4-9. Voltage notching is a switching disturbance of the normal voltage waveform lasting less than half a cycle.

Voltage Unbalance

While voltage unbalance at motor terminals can adversely affect motor operation, it can also cause problems at the line side of the drive. As little as 0.3% voltage unbalance on the input to a PWM inverter can cause voltage notching and excessive current to flow in one or more phases. This can cause tripping of the VSD's current overload fault protection.

A common cause of voltage unbalance is single-phase loads dropping in or out on the same feed as the three-phase VSD. Increasing the kVA rating of the transformer or providing a separate feed for the VSD can help minimize or eliminate this type of problem.

TECH TIP
For every 1% voltage unbalance, there can be up to a 6% current unbalance. The greater the voltage or current unbalance, the higher the operating temperature of the motor.

Harmonics

VSDs are affected by harmonics. VSDs also create harmonic problems for other loads in the system. There are several global standards (IEEE, ISO, etc.) that detail the acceptable levels of harmonics that may be produced by a drive in a typical operating environment.

VSDs with Harmonics. Voltage flat-topping is a problem that can occur when installing a VSD in an environment with harmonics. During the rectification of AC to DC, current is drawn only at the peak of the AC voltage waveform, when the AC line voltage exceeds the voltage of the DC bus capacitor.

The burst of current into the capacitor at the peak of the AC voltage waveform can cause voltage to drop. If there is flat-topping of the line voltage caused by other electronic loads in the building, the DC bus capacitor in the VSD cannot charge to full capacity. This can cause drops in the DC bus voltage if the motor load suddenly increases or if there is a sudden sag in the line voltage. Increasing the transformer kVA rating and distribution conductors' capacity or adding voltage regulation, such as a UPS, can correct this problem.

VSD Harmonics Creation. High power VSDs (or many low-power VSDs, an increasingly common situation) can create harmonics that are large enough to cause problems for other loads in the system. If the VSD installation exceeds the capacity of the building's distribution transformer and/or distribution wiring, flat-topping may occur and cause problems for other loads in the facility. Harmonic currents are only problematic if they encounter source impedance. If the utility transformer has too little VA capacity for the level of harmonic currents present, the source voltage begins to distort.

Although traditional electric motors can fail for several reasons, causes of electric motor failure with modern equipment include the following:

- Electric motors with modern equipment can fail due to downsizing to minimum-rated motors (in power and amperage) to save energy. Using minimum-rated motors can increase motor wear and result in more maintenance and downtime.

- Use of modern technologies such as motor drives, programmable logic controllers (PLCs), and other solid-state controls introduces new problems. These problems include harmonics (overtones resulting from a frequency that is a multiple of the fundamental frequency) and transient spikes in the system. Such problems can cause motors and other electrical components to become damaged or fail.

Motor Drive Problems

Motor drives are considered to be a modern type of motor control when compared with traditional motor control with magnetic motor starters, so the reasons that they fail have not been as thoroughly documented. Common causes for motor drive failure include the following:

- All motor drives must be programmed and reprogrammed so that their operating parameters match the motor and motor load type. There are several range settings that if not correctly programmed can damage the motor and drive and cause an unsafe operating condition.

- In order to save energy, drives and motors are sized for the minimum size that can perform the required work. Downsizing equipment saves energy, but forces the drive and motor to operate at almost full capacity.

- Electronic devices are susceptible to problems caused by overheating. If problems from overheating are not addressed by installing heat sinks, ventilation, or cooling fans, electronic devices and circuits inside motor drives can become damaged.

- The increased use of nonlinear loads such as computers, printers, and motor drives introduces more harmonics on the power distribution system, which can cause overheating and related problems.

- As electronic circuits are reduced in size and the devices are placed closer together, even small transient voltages can cause arcing and damage components and devices.

CASE STUDY: Discrete Parts Manufacturing System

Discrete parts manufacturing is the production of durable goods that are produced and assembled such as automobiles, refrigerators, computers, tools, toys, and other tangible items that are expected to be in use for time periods of up to several years. Discrete parts manufacturing is performed primarily by process control equipment that bends, punches, grinds, stamps, forms, cleans, paints, assembles, and performs quality control on the product produced.

In most discrete parts manufacturing systems, one piece of equipment or process feeds into another as the product is assembled or manufactured. Each electric motor along the production line is typically controlled by a separate motor drive, but certain motor drives are used to control several motors. Although installing only one motor drive can simplify process control, it can also produce problems such as differences in voltage and current. **See Figure 4-10.**

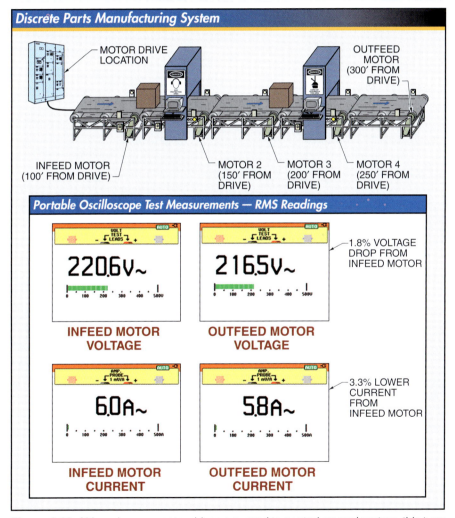

Figure 4-10. Although some motor drives are used to control several motors, this type of configuration can produce problems such as differences in voltage and current.

The Problem

In a discrete parts manufacturing application, the outfeed motor has failed twice in the last year and was replaced after each failure. The following steps were taken after the motor was replaced:

1. Voltage measurements were taken at the infeed and outfeed motors to verify the longer distance was not creating a voltage sag condition. The numerical measurements indicated only a 1.8% voltage drop, which is within the acceptable range.

2. Current measurements were taken at the infeed and outfeed motors to verify there was not an overcurrent problem. Both current measurements were within the range of motor nameplate specifications.

3. Although the numerical readings from the initial measurements did not indicate a problem, additional measurements were taken using a portable oscilloscope. The portable oscilloscope voltage measurements made at the outfeed motor did not indicate the presence of transient voltages that could be caused by the longer distance, but the current measurement showed an increased feed current.

4. Load reactors were installed before the outfeed motor, and the portable oscilloscope indicated that they were properly operating.

Problem Analysis

When the portable oscilloscope waveform for current was analyzed, it indicated a 26% increase in the size of the peak current readings. The increase in peak current was determined to be the main cause of motor failure. **See Figure 4-11.**

Problem Solutions

Changes and solutions that can be implemented to a discrete manufacturing parts system to solve outfeed motor failure include the following:

• The size and ratings of the outfeed motor can be increased to be within the range of the increase in peak current.

• An improved solution is to relocate the main motor drive to the center location of the process or equipment to reduce the distance between the motor drive and the motor.

• The preferred solution is to reduce the distance between the motor drive and the motors, with each motor controlled by a separate motor drive. Multiple motor drives can be controlled by a common control so that they operate as one unit. **See Figure 4-12.**

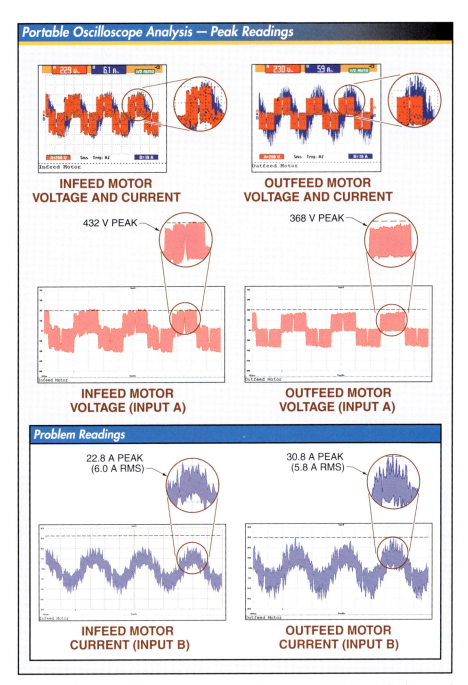

Figure 4-11. Portable oscilloscopes can be used to analyze waveforms of voltage and current and detect previously unseen problems.

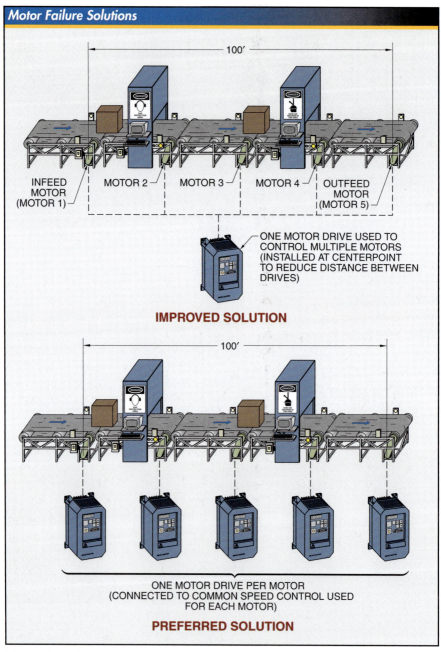

Motor Failure Solutions

100′

INFEED MOTOR (MOTOR 1) — MOTOR 2 — MOTOR 3 — MOTOR 4 — OUTFEED MOTOR (MOTOR 5)

ONE MOTOR DRIVE USED TO CONTROL MULTIPLE MOTORS (INSTALLED AT CENTERPOINT TO REDUCE DISTANCE BETWEEN DRIVES)

IMPROVED SOLUTION

100′

ONE MOTOR DRIVE PER MOTOR (CONNECTED TO COMMON SPEED CONTROL USED FOR EACH MOTOR)

PREFERRED SOLUTION

Figure 4-12. Separate motor drives are the preferred method of motor control and prevention of motor failure when several motors are installed in a discrete parts manufacturing system.

5

Documenting Motor and Drive Condition

Documenting motor and drive condition and maintaining accurate records of downtime, maintenance requests, equipment condition, and troubleshooting provides verification and justification for requirements concerning personnel, overtime, tools, test instruments, and training. Recording and documenting specific measurements while performing troubleshooting operations provides written information about equipment and circuit operation that can be reviewed when troubleshooting, determining possible corrective action, making queries to the manufacturer, and selecting replacement equipment.

DOCUMENTATION

Documenting measured test results with a portable oscilloscope when installing new equipment, conducting preventive maintenance, and troubleshooting requires knowledge of the test instrument. This knowledge includes knowing how to select the proper scope function, set the test instrument to record the proper measurements, record the data over time, replay the data, analyze the recorded measurements, and apply the measurements to specific equipment and systems. **See Figure 5-1.**

TECH TIP

An archive of waveforms and other measurements taken provides a valuable baseline that enables a technician to make comparisons of equipment performance over time.

Recording and Analyzing Measurements over Time

In order to take and record measurements over time, portable oscilloscopes must have a time measurement function. The time measurement function can be used to record both voltage and current over a specified time period.

Once electrical quantities have been displayed and recorded over time, they can be analyzed. A cursor can be displayed on the recorded screen and moved over any section of the recorded measurements to verify what the test measurements are and the point in time in which they were recorded. For example, the cursor position indicates that the voltage is 120 V and current is 12.1 A prior to the spike in current. As the cursor is moved over the event, the voltage dips from 120 V to 115 V and current increases from 12.1 A to 58.5 A. **See Figure 5-2.**

Portable Oscilloscope Measurement Documentation

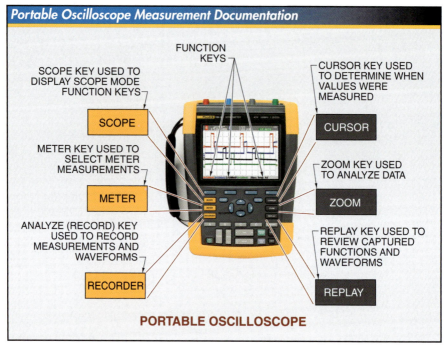

FUNCTION
KEYS

SCOPE KEY USED TO
DISPLAY SCOPE MODE
FUNCTION KEYS

SCOPE

METER KEY USED TO
SELECT METER
MEASUREMENTS

METER

ANALYZE (RECORD) KEY
USED TO RECORD
MEASUREMENTS AND
WAVEFORMS

RECORDER

CURSOR KEY USED
TO DETERMINE WHEN
VALUES WERE
MEASURED

CURSOR

ZOOM KEY USED
TO ANALYZE DATA

ZOOM

REPLAY KEY USED TO
REVIEW CAPTURED
FUNCTIONS AND
WAVEFORMS

REPLAY

PORTABLE OSCILLOSCOPE

Figure 5-1. The SCOPE, METER, RECORDER, CURSOR, ZOOM, and REPLAY buttons on a portable oscilloscope are used to set the type of measurement taken and to record, document, and analyze readings.

Use a power quality analyzer to assess the quality of the power supply to the panel.

Recording and Analyzing Waveforms

Recording and analyzing numerical measurements can be performed to help understand how the circuit operates. It is also helpful to record, view, and analyze waveforms. For example, a portable oscilloscope can be programmed to record the waveforms from the startup and operation of a motor and motor drive to analyze the effects it has on the system. **See Figure 5-3.** As waveforms are captured and displayed, the display records the circuit transition from a linear circuit to a nonlinear circuit as the motor drive turns on the motor.

Recording and Analyzing Measurements over Time

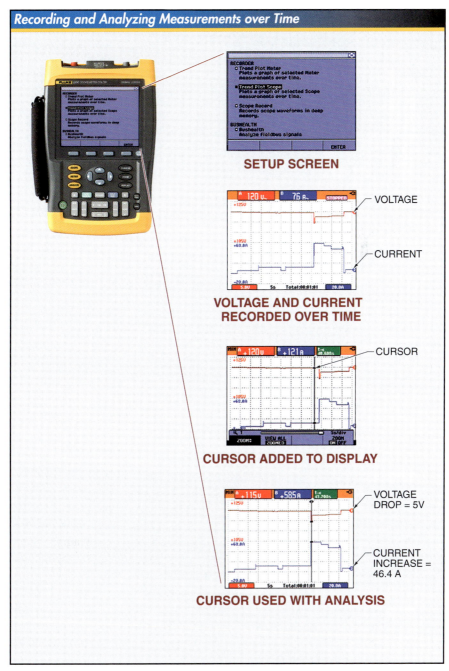

SETUP SCREEN

VOLTAGE

CURRENT

**VOLTAGE AND CURRENT
RECORDED OVER TIME**

CURSOR

CURSOR ADDED TO DISPLAY

VOLTAGE
DROP = 5V

CURRENT
INCREASE =
46.4 A

CURSOR USED WITH ANALYSIS

Figure 5-2. The time measurement function on a portable oscilloscope can be used to record both voltage and current over a specified time period.

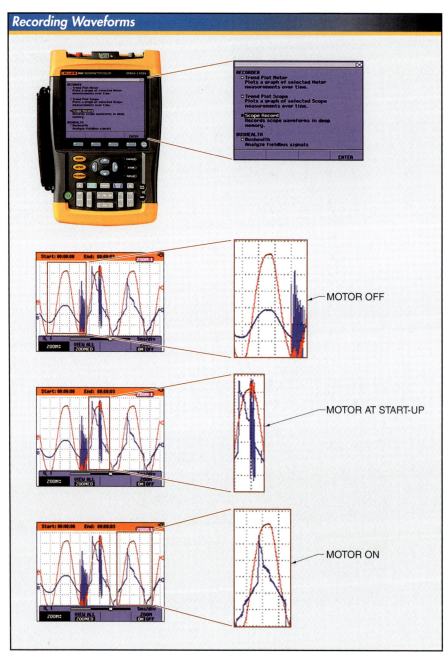

Figure 5-3. A portable oscilloscope can be set up to record waveforms from the startup and continuous operation of a motor and motor drive to check that it is operating properly.

TECH TIP

Time delay fuses and circuit breakers are used in motor circuits because all motors draw higher currents when starting (inrush current) than when running. Only a portable oscilloscope can capture and display the actual motor inrush current for analyzing problems during starting.

Adding the cursor to the display screen verifies that the linear current

was 8.2 A before the motor was turned on, increased to 43 A at motor startup, and was constant at 30.6 A while the motor was running. This information is required for understanding fuse and circuit breaker trip problems, damage to equipment, and other information that can be used to maintain and troubleshoot equipment and system problems. **See Figure 5-4.**

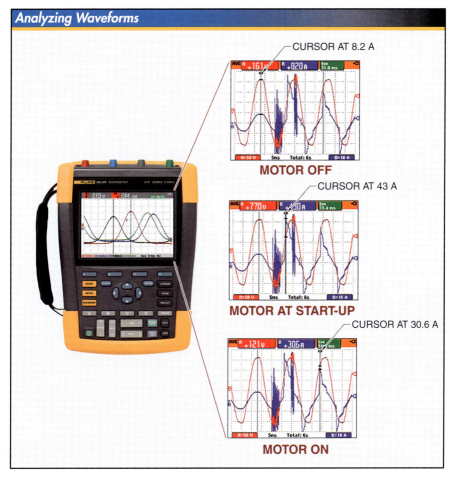

Figure 5-4. Adding a cursor to the display screen helps analyze the waveforms from the recorded measurements.

Waveform Shapes. More information can be determined from the shape of a waveform than a numerical value. Understanding what is displayed and the actual measurement value is required when viewing waveforms displayed on the portable oscilloscope screen. To help understand the values and shapes of a displayed waveform, the crest factor must be properly analyzed. A *crest factor* is the ratio of the peak voltage value to the rms voltage value. A pure sinusoidal waveform has a 1.41 crest factor. The higher the crest factor in a circuit, the more distorted the waveform. **See Figure 5-5.**

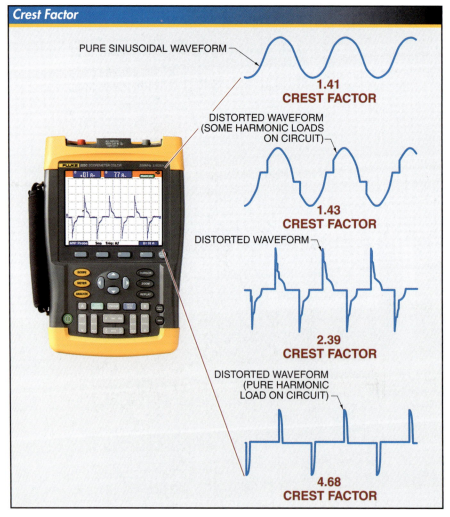

Crest Factor

PURE SINUSOIDAL WAVEFORM

1.41 CREST FACTOR

DISTORTED WAVEFORM (SOME HARMONIC LOADS ON CIRCUIT)

1.43 CREST FACTOR

DISTORTED WAVEFORM

2.39 CREST FACTOR

DISTORTED WAVEFORM (PURE HARMONIC LOAD ON CIRCUIT)

4.68 CREST FACTOR

Figure 5-5. A pure sinusoidal waveform has a 1.41 crest factor. The higher the crest factor in a circuit the more distorted the waveform.

CASE STUDY: Recording and Analyzing Measurements

Once test measurements are recorded and analyzed, proper corrective action can be taken. For example, in a variable torque application, a portable oscilloscope set to pumps and fans (P&F) was used to record, analyze, and document problems such as voltage sags and swells, power interruptions, transients, overcurrents, poor power factor, line noise, and harmonics. The information captured and displayed clearly indicated that one of these problems is present. Based on the documentation from the portable oscilloscope, it was determined that a line filter needed to be placed between the drive and motors to correct the problem.

Once a specific problem is identified, the proper corrective action can be taken. After the corrective action is taken, additional measurements with waveforms can be taken to verify that the corrective action solved the problem. **See Figure 5-6.**

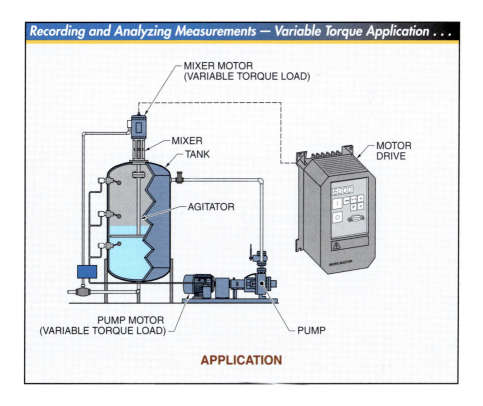

Recording and Analyzing Measurements — Variable Torque Application . . .

MIXER MOTOR
(VARIABLE TORQUE LOAD)

MIXER

TANK

MOTOR
DRIVE

AGITATOR

PUMP MOTOR
(VARIABLE TORQUE LOAD)

PUMP

APPLICATION

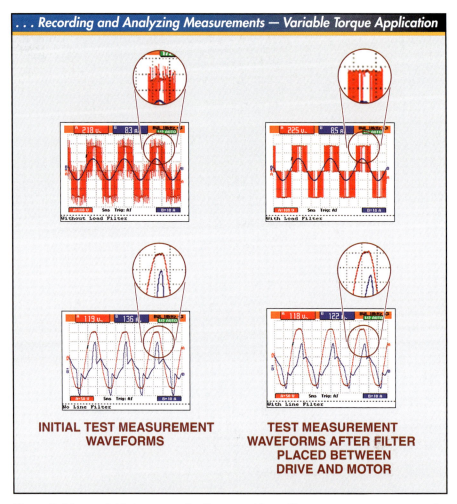

. . . *Recording and Analyzing Measurements — Variable Torque Application*

INITIAL TEST MEASUREMENT WAVEFORMS

TEST MEASUREMENT WAVEFORMS AFTER FILTER PLACED BETWEEN DRIVE AND MOTOR

Figure 5-6. In a variable torque application, portable oscilloscope measurements can be used to record, analyze, and document problems such as voltage sags and swells, power interruptions, transients, overcurrents, poor power factor, line noise, and harmonics.

Electric Motor Drive Component Failures

Symptom/Fault Code	Problem	Cause	Solution
Electric motor drive does not turn on; blown fuse or tripped circuit breaker	Defective converter (rectifier) semiconductor	High input voltage/voltage swell	Replace converter semiconductor or replace electric motor drive
		Electric motor drive cooling fan is defective	Replace converter semiconductor and electric motor drive cooling fan, or replace electric motor drive
Electric motor drive does not turn on; blown fuse or tripped circuit breaker; no AC output from electric motor drive; DC bus fuse (if present) is blown; electric motor drive overcurrent fault	Defective inverter semiconductor	Electric motor drive cooling fan is defective	Replace inverter semiconductor and electric motor drive cooling fan, or replace electric motor drive
Electric motor drive overtemperature fault or electric motor drive component failure	Defective electric motor drive cooling fan	Age; electric motor drive manufacturers recommend replacing electric motor drive cooling fans every 3 to 5 years; *Note: Fan life expectancy is influenced by ambient temperature*	Replace electric motor drive cooling fan or replace electric motor drive
DC bus capacitor is swollen and/or pressure relief valve is protruding; DC bus capacitor is destroyed; unstable DC bus voltage when electric motor drive is running at a constant speed; *Note: Defective capacitor balancing resistors can damage the DC bus capacitors; if there is a problem with a DC bus capacitor, also check the capacitor balancing resistor*	Defective DC bus capacitors	Age; manufacturers recommend replacing capacitors every 5 to 10 years; electric motor drive cooling fan is defective.	Replace DC bus capacitors or replace electric motor drive
		Capacitor balancing resistors are defective or incorrect value	Replace DC bus capacitors and electric motor drive cooling fan, or replace electric motor drive
			Replace DC bus capacitors and capacitor balancing resistors, or replace electric motor drive; verify that replacement resistors are correct ohm and watt value
Resistors and resistor connections are discolored and/or burned; *Note: Defective capacitor balancing resistors can damage the DC bus capacitors; if there is a problem with a capacitor balancing resistor, also check the DC bus capacitor*	Defective capacitor balancing resistors	Incorrect capacitor balancing resistors	Replace capacitor balancing resistors, or replace electric motor drive; verify that replacement resistors are correct ohm and watt value
		Excessive heat	Verify that electric motor drive cooling fan is working, replace if defective; replace capacitor balancing resistors, or replace electric motor drive; verify that replacement resistors are correct ohm and watt value
		Loose connections	Replace capacitor balancing resistors, or replace electric motor drive; verify that replacement resistors are correct ohm and watt value; tighten connections

AC Motor Characteristics

Motor Type 1φ	Typical Voltage	Starting Ability (Torque)	Size (HP)	Speed Range (rpm)	Cost*	Typical Uses
Shaded-pole	115 V, 230 V	Very low, 50% to 100% of full load	Fractional ½ HP to ⅓ HP	Fixed 900,1200, 1800, 3600	Very low, 75% to 85%	Light-duty applications such as small fans, hair dryers, blowers, and computers
Split-phase	115 V, 230 V	Low, 75% to 200% of full load	Fractional ⅓ HP or less	Fixed 900,1200, 1800, 3600	Low, 85% to 95%	Light-torque applications such as pumps, blowers, fans, and machine tools
Capacitor-start	115 V, 230 V	High, 200% to 350% of full load	Fractional to 3 HP	Fixed 900,1200, 1800	Low, 90% to 110%	Hard-to-start loads such as refrigerators, air compressors, and power tools
Capacitor-run	115 V, 230 V	Very low, 50% to 100% of full load	Fractional to 5 HP	Fixed 900,1200, 1800	Low, 90% to 110%	Applications that require a high running torque such as pumps and conveyors
Capacitor start-and-run	115 V, 230 V	Very low, 350% to 450% of full load	Fractional to 10 HP	Fixed 900,1200, 1800	Low, 100% to 115%	Applications that require a high starting and running torque such as loaded conveyors
3φ Induction	230 V, 460 V	Low, 100% to 175% of full load	Fractional to over 500 HP	Fixed 900,1200, 3600	Low, 100%	Most industrial applications
Wound rotor	230 V, 460 V	High, 200% to 300% of full load	½ HP to 200 HP	Varies by changing resistance in rotor	Very high, 250% to 350%	Applications that require high torque at different speeds such as cranes and elevators
Synchronous	230 V, 460 V	Very low, 40% to 100% of full load	Fractional to 250 HP	Exact constant speed	High, 200% to 250%	Applications that require very slow speeds and correct power factors

* based on standard 3φ induction motor

DC and Universal Motor Characteristics

Motor Type	Typical Voltage	Starting Ability (Torque)	Size (HP)	Speed Range (rpm)	Cost*	Typical Uses
DC Series	12 V, 90 V, 120 V, 180 V	Very high, 400% to 50% of full load	Fractional to 100 HP	Varies 0 to full speed	High 175% to 225%	Applications that require very high torque such as hoists and bridges
Shunt	12 V, 90 V, 120 V, 180 V	Low, 125% to 250% of full load	Fractional to 100 HP	Fixed or adjustable below full speed	High 175% to 225%	Applications that require better speed control than a series motor such as woodworking machines
Compound	12 V, 90 V, 120 V, 180 V	High, 300% to 400% of full load	Fractional to 100 HP	Fixed or adjustable	High 175% to 225%	Applications that require high torque and speed control such as printing presses, conveyors, and hoists
Permanent-magnet	12 V, 24 V, 36 V, 120 V	Low, 100% to 200% of full load	Fractional	Varies from 0 to full speed	High 150% to 200%	Applications that require small DC-operated equipment such as automobile power windows, seats, and sun roofs
Stepping	5 V, 12 V, 24 V	Very low† 0.5 to 5000 oz/in.	Size rating is given as holding torque and number of steps	Rated in number of steps per sec (maximum)	Varies based on number of steps and rated torque	Applications that require low torque and precise control such as indexing tables and printers
AC/DC universal	115 V, 230 VAC, 12 VDC, 24 VDC, 36 VDC, 20 VDC	High, 300% to 400% of full load	Fractional	Varies 0 to full speed	High 175% to 225%	Most portable tools such as drills, routers, mixers, and vacuum cleaners

* based on standard 3φ induction motor
† torque is rated as holding torque

Electric Motor Drive Troubleshooting Matrix

Electric Motor Drive Faults

Symptom/Fault Code	Problem	Cause	Solution
Electric motor drive overvoltage fault	Electric motor drive overvoltage	Deceleration time is too short	Increase deceleration time
		High input voltage/ voltage swell	Correct as needed
		Load is overhauling motor	Add dynamic braking resistors and/or increase deceleration time
Electric motor drive overcurrent fault	Electric motor drive overcurrent	Motor nameplate data incorrectly programmed	Check motor nameplate and program data correctly
		Acceleration time is too short	Increase acceleration time
		Start boost or continuous boost set too high	Lower start boost or continuous boost
		Short in inverter semiconductor	Replace inverter semiconductor or replace electric motor drive
		Contactor between electric motor drive and motor is changing state while electric motor drive outputs more than 0 Hz	Wire contactor to change state only when electric motor drive outputs 0 Hz
		Motor trying to start in a spinning mode	Enable flying start
Electric motor drive over-current fault; load conductor current readings not equal	Electric motor drive overcurrent	Motor or conductors feeding motor are shorted or have ground fault	Correct as needed
Electric motor drive over-current fault; current readings for load conductors are 105% or greater than motor nameplate current when motor powers driven load	Electric motor drive overcurrent	Problem with motor	
Electric motor drive over-load fault; current readings for load conductors are 105% or greater than motor nameplate current when motor powers driven load	Electric motor drive overload	Problem with motor and/or load	Correct as needed
Electric motor drive undervoltage fault	Electric motor drive undervoltage	Low input voltage/ voltage sag	
Electric motor drive overtemperature fault	Electric motor drive overtemperature	High ambient temperature	Add cooling to electric motor drive enclosure, or relocate electric motor drive enclosure
		Drive cooling fan defective	Replace electric motor drive cooling fan, or replace electric motor drive
		Heat sink dirty, or air intake clogged	
		Problem with motor and/or load	Clean heat sink or air intake

Electric Motor Drive Parameter Problems

Symptom/Fault Code	Problem	Cause	Solution
Electric motor drive operates correctly when set to default parameters, input mode is keypad, and motor/load is disconnected. Electric motor drive does not operate correctly when set to default parameters (excluding motor nameplate data and control mode), input mode is keypad, and motor/load is connected	Parameters incorrect	Parameter(s) incorrectly programmed	Verify that the motor nameplate data parameter matches the actual motor nameplate data; verify that the control mode parameter is correct for the type of load, e.g., electric motor drive set to variable torque for a pump or fan load Correct as needed
		Parameter(s) incorrectly programmed	Verify that all parameters that pertain to application are set correctly; it is possible there is a conflict between two parameters, e.g., minimum motor frequency and maximum motor frequency
	Problem with motor and/or load	Problem with motor and/or load	Correct as needed
Unusual noises or vibrations when electric motor drive is powering the load	Parameter(s) incorrect	Parameter(s) incorrectly programmed	Adjust skip frequency parameter
	Problem with motor and/or load	Problem with motor and/or load	Correct as needed
Electric motor drive operates correctly when set to default parameters (excluding motor nameplate data and control mode), input mode is keypad, and motor/load is connected; electric motor drive does not operate correctly when input mode is other than keypad, motor/load is connected, and electric motor drive is operated per design	Parameter(s) incorrect	Parameter(s) incorrectly programmed	Verify that input mode parameter matches the existing input, e.g., serial communication is programmed if electric motor drive is controlled via serial communication Correct as needed
	Externally connected inputs and outputs incorrect	Input(s) and/or output(s) incorrectly wired; input or ouput devices not correct.	See *Electric Motor Drive Input and Output Problems*

Electric Motor Drive Operational Problems

Symptom/Fault Code	Problem	Cause	Solution
Intermittent electric motor drive faults and/or intermittently electric motor drive does not operate as per design	Intermittent fault/erratic operation	Incoming power problems	Correct as needed
		Electrical noise, EMI/RFI	Verify RC snubbers, MOVs, or flywheel diodes are installed across coils operating near electric motor drive or controlled by electric motor drive; correct as needed.
		Electrical noise, EMI/RFI	Verify that input and output conductors are installed in separate metal raceways or separate shielded cables; verify proper grounding; correct as needed
		Electrical noise, EMI/RFI	Verify that proper separation is maintained between power and control conductors, including separate metal raceways; correct as needed
		Electrical noise, EMI/RFI	Verify that analog signals are run in a separate metal conduit using shielded twisted pair cable, and only grounded at one end; correct as needed
Motor rotation incorrect when powered by electric motor drive	Incorrect phasing	Wiring	Interchange two of the load conductors at the power terminal strip to reverse the direction of rotation
Motor rotation incorrect when powered through bypass contactors	Incorrect phasing	Wiring	Interchange two of the line conductors that feed the electric motor drive and the bypass; Note: This assumes that the electric motor drive and bypass share a common feed

Electric Motor Drive Input and Output Problems

Symptom/Fault Code	Problem	Cause	Solution
Electric motor drive operates correctly when set to default parameters (excluding motor nameplate data and control mode), input mode is keypad, and motor/load is connected; electric motor drive does not operate correctly when input mode is other than keypad, motor/load is connected, and electric motor drive is operated per design	Externally connected inputs and outputs incorrect	Input(s) and/or output(s) incorrectly wired; input or output devices not correct	Verify the wiring to input and output devices; check for loose connections; verify the devices are correct for the application; correct as needed
		Problem with separate system that supplies start, stop, reference, or feedback signals to the electric motor drive, e.g., HVAC control system	Verify the operation of separate system; correct as needed
	Parameter(s) incorrect	Parameter(s) incorrectly programmed	See *Electric Motor Drive Parameter Problems.*

3φ, 230 V Motors and Circuits — 240 V System

1		2		3	4	5				6	
Size of motor		Motor overload protection — Low-peak or Fusetron®				Controller termination temperature rating				Minimum size of copper wire and trade conduit	
						60°C		75°C			
HP	Amp	Motor less than 40°C or greater than 1.15 SF (Max fuse 125%)	All other motors (Max fuse 115%)	Switch 115% minimum or HP rated or fuse holder size	Minimum size of starter	TW	THW	TW	THW	Wire size (AWG or kcmil)	Conduit (inches)
½	2	2½	2¼	30	00	•	•	•	•	14	½
¾	2.8	3½	$3\frac{2}{10}$	30	00	•	•	•	•	14	½
1	3.6	4½	4	30	00	•	•	•	•	14	½
1½	5.2	6¼	$5\frac{6}{10}$	30	00	•	•	•	•	14	½
2	6.8	8	7½	30	0	•	•	•	•	14	½
3	9.6	12	10	30	0	•	•	•	•	14	½
5	15.2	17½	17½	30	1	•	•	•	•	14	½
7½	22	25	25	30	1	•	•	•	•	10	½
10	28	35	30	60	2	•	•	•		8	¾
									•	10	½
15	42	50	45	60	2	•	•	•	•	6	1
										6	¾
20	54	60	60	100	3	•	•	•	•	4	1
25	68	80	75	100	3	•	•			3	1¼
								•		3	1
									•	4	1
30	80	100	90	100	3	•	•	•		1	1¼
									•	3	1¼
40	104	125	110	200	4	•	•	•		2/0	1½
									•	1	1¼
50	130	150	150	200	4	•	•	•		3/0	2
									•	2/0	1½
75	192	225	200	400	5	•	•	•		300	2½
									•	250	2½
100	248	300	250	400	5	•	•	•		500	3
									•	350	2½
150	360	450	400	600	6	•	•	•		300-2/φ*	2-2½*
									•	4/0-2/φ*	2-2*

* two sets of multiple conductors and two runs of conduit required

3φ , 460 V Motors and Circuits — 480 V System

1		2		3	4	5				6	
Size of motor		Motor overload protection		Switch 115% minimum or HP rated or fuse holder size	Minimum size of starter	Controller termination temperature rating				Copper wire and trade conduit	
		Low-peak or Fusetron®				60°C		75°C			
		Motor less than 40°C or greater than 1.15 SF (Max fuse 125%)	All other motors (Max fuse 115%)			TW	THW	TW	THW	Wire size (AWG or kcmil)	Conduit (inches)
HP	Amp										
½	1	1¼	1⅛	30	00	•	•	•	•	14	½
¾	1.4	1⁶/₁₀	1⁶/₁₀	30	00	•	•	•	•	14	½
1	1.8	2¼	2	30	00	•	•	•	•	14	½
1½	2.6	3²/₁₀	2⁸/₁₀	30	00	•	•	•	•	14	½
2	3.4	4	3½	30	00	•	•	•	•	14	½
3	4.8	5⁶/₁₀	5	30	0	•	•	•	•	14	½
5	7.6	9	8	30	0	•	•	•	•	14	½
7½	11	12	12	30	1	•	•	•	•	14	½
10	14	17½	15	30	1	•	•	•	•	14	½
15	21	25	20	30	2	•	•	•	•	10	½
20	27	30	30	60	2	•	•	•		8	¾
									•	10	½
25	34	40	35	60	2	•	•	•		6	1
									•	8	¾
30	40	50	45	60	3	•	•	•		6	1
									•	8	¾
40	52	60	60	100	3	•	•	•		4	1
									•	6	1
50	65	80	70	100	3	•	•	•		3	1¼
									•	4	1
60	77	90	80	100	4	•	•	•		1	1¼
									•	3	1¼
75	96	110	110	200	4	•	•	•		1/0	1½
									•	1	1¼
100	124	150	125	200	4	•	•	•		3/0	2
									•	2/0	1½
125	156	175	175	200	5	•	•	•		4/0	2
									•	3/0	2
150	180	225	200	400	5	•	•	•		300	2½
									•	4/0	2
200	240	300	250	400	5	•	•	•		500	3
									•	350	2½
250	302	350	325	400	6	•	•	•		4/0-2/φ*	2-2*
									•	3/0-2/φ*	2-2*
300	361	450	400	600	6	•	•	•		300-2/φ*	2-1½ *
									•	4/0-2/φ*	2-2*

* two sets of multiple conductors and two runs of conduit required

A

alternating current (AC) motor: A motor that uses alternating current connected to a stator (stationary windings) to produce a force on a rotor through a magnetic field.

alternation: Half of a cycle (180°).

apparent power (P_A): The product of voltage and current in a circuit calculated without considering the phase shift that may be present between the voltage and current in a circuit; measured in volt amps (VA) or kilovolt amps (kVA).

average voltage (V_{avg}): The mathematical mean of all instantaneous voltages of a half cycle of a sine wave.

B

bandwidth: The range of frequencies that a portable oscilloscope or meter can accurately measure.

breakdown torque (BDT): The maximum torque a motor can provide without an abrupt reduction in motor speed.

C

carrier frequency: The frequency that controls the rate at which solid-state switches in the inverter of a pulse width modulated (PWM) motor drive turn on and off.

constant horsepower (CH) load: A load that requires high torque at low speeds and low torque at high speeds.

constant torque (CT) load: A load in which the motor torque requirement remains constant.

crest factor: The ratio of the peak voltage value to the rms voltage value.

current: The flow of electrons in an electrical circuit; measured in amperes (A).

D

digital multimeter (DMM): An electrical testing instrument that measures two or more electrical properties and displays the measured properties as numerical values.

digital thermometer: A device used to take temperature measurements on energized circuits or on moving parts without contacting the point of measurement and provide measurement readings on a digital display.

diode bridge: An electrical circuit that is used to convert AC into DC. It contains four diodes that permit both halves of input alternating current sine waves to pass.

direct current (DC) motor: A motor that uses direct current connected to the field windings and rotating armature to produce rotation.

dynamic braking: A method of motor braking in which a motor is reconnected to act as a generator immediately after it is turned off.

E

electric motor: A machine that converts electrical power into rotating mechanical force (torque) on a shaft, which is used to produce work.

energy efficiency (green technology): The process of selecting, replacing, or modifying equipment and systems to reduce the amount of power used and still provide the same amount of service.

F

full-load torque (FLT): The torque required to produce the rated power at the full speed of a motor.

fundamental frequency: The frequency of the voltage used to control motor speed.

H

hertz (Hz): The international unit of frequency; equal to cycles per second.

I

insulated gate bipolar transistor (IGBT): A solid-state, three-terminal device that is used for fast switching of electric circuits.

insulation resistance tester: A test instrument that has specific functions for insulation resistance testing.

International Electrotechnical Commission (IEC): An organization that develops international standards for electrical equipment.

L

linear load: A load in which current increases proportionately as the voltage increases, and current decreases proportionately as voltage decreases.

line frequency: The amount of complete electrical cycles per second of a power source.

locked rotor torque (LRT): The torque a motor produces when its rotor is stationary and full power is applied to the motor.

lockout: The process of removing the source of electrical power and installing a lock that prevents the power from being turned on.

M

motor drive: An electronic unit designed to control the speed of a motor using solid-state devices.

motor torque: The force that produces rotation in a motor.

N

nonlinear load: A load in which the instantaneous load current is not proportional to the instantaneous voltage.

P

peak-to-peak voltage (V_{p-p}): The value measured from the maximum positive alternation to the maximum negative alternation.

peak voltage (V_{max}): In a sine wave, the maximum value of either the positive or negative alternation.

personal protective equipment (PPE): Clothing and/or equipment worn by a person to reduce the possibility of injury in the work area.

polarity: The positive (+) or negative (−) electrical state of an object.

portable oscilloscope (ScopeMeter™): A test instrument that measures and displays the waveforms of high-voltage power, low-voltage control, and digital signals.

power factor (PF): The ratio of true power used in an AC circuit to apparent power used by or delivered to the circuit.

power quality meter: A test instrument that measures, displays, and records voltage, current, and power in addition to special power problems such as sags, swells, transients, and harmonics.

preventive maintenance: Scheduled work required to keep equipment in peak operating condition.

pull-up torque (PUT): The torque required to bring a load up to its rated speed.

pulse width modulation (PWM): A method of controlling the amount of voltage sent to the motor.

Q

qualified person: A person who is trained and has special knowledge of the construction and operation of electrical equipment or a specific task and is trained to recognize and avoid electrical hazards that might be present with respect to the equipment or specific task.

R

reactive power (VAR): Power supplied to reactive loads, such as motor coil windings.

root-mean-square (effective) voltage (V_{rms}): The voltage that produces the same amount of heat in a pure resistive circuit as is produced by an equal DC voltage.

S

sampling: The process of converting a portion of an input signal into a number of discrete electrical values for the purpose of storage, processing, and display.

service factor amperage (SFA): The maximum current rating a motor can safely draw.

sweep: The movement of the displayed trace across an oscilloscope screen.

T

tagout: The process of placing a danger tag on the source of electrical power, which indicates that the equipment may not be operated until the danger tag is removed.

thermal imager: A device that detects heat patterns in the infrared wavelength spectrum without making direct contact with the workpiece.

torque: The force that produces rotation.

trace: A line that sweeps across the display screen of a portable oscilloscope to display a signal's amplitude variations over time.

transient voltage: An undesirable, momentary voltage pulse that varies in amplitude and energy level depending upon the source (solenoid coil or lightning strike) of the transient.

troubleshooting: The systematic elimination of the various parts of a system or process from consideration during the process of locating a malfunctioning part.

troubleshooting procedure: A logical step-by-step process used to identify a malfunction or problem in a system or process as quickly and easily as possible.

true power (P_T): The actual power used in an electrical circuit; measured in watts (W) or kilowatts (kW).

V

variable torque (VT) load: A load that requires a varying torque and horsepower at different speeds.

voltage: The amount of electrical "pressure" in a circuit; measured in volts (V) and can be either direct current (DC) or alternating current (AC).

voltage notch: A fast switching disturbance of the normal voltage waveform.

volts/hertz (V/Hz) ratio: The ratio of the voltage and frequency applied to a motor.

Index